MW00479618

Son of Apollo

Christopher A. Roosa

www.roosa.com

Outward Odyssey
A People's History of Spaceflight

Series editor
Colin Burgess

Son of Apollo

The Adventures of a Boy Whose Father Went to the Moon

Christopher A. Roosa

Foreword by Jim Lovell

UNIVERSITY OF NEBRASKA PRESS • LINCOLN

The University of Nebraska Press is part of
a land-grant institution with campuses and
programs on the past, present, and future
homelands of the Pawnee, Ponca, Otoe-
Missouria, Omaha, Dakota, Lakota, Kaw,
Cheyenne, and Arapaho Peoples, as well as
those of the relocated Ho-Chunk, Sac and
Fox, and Iowa Peoples.

Library of Congress Cataloging-in-
Publication Data
Names: Roosa, Christopher A., author. |
Lovell, Jim, writer of foreword.
Title: Son of Apollo: the adventures of a boy
whose father went to the moon / Christopher
A. Roosa; foreword by Jim Lovell.
Description: Lincoln: University of Nebraska
Press, 2022. | Series: Outward odyssey: a
people's history of spaceflight.
Identifiers: LCCN 2022008550
ISBN 9781496233349 (hardback)
ISBN 9781496234230 (epub)
ISBN 9781496234247 (pdf)
Subjects: LCSH: Roosa, Christopher A. |
Roosa, Stuart A. (Stuart Allen), 1933–1994—
Family. | Project Apollo (U.S.) | United
States. National Aeronautics and Space
Administration—History. | Children of
astronauts—United States—Biography. |
Fathers and sons—United States. | Space
flight to the moon—History. | BISAC:
BIOGRAPHY & AUTOBIOGRAPHY /
Adventurers & Explorers
Classification: LCC TL789.85.R655 A3 2022 |
DDC 629.450092 [B]—dc23/eng/20220624
LC record available at
https://lccn.loc.gov/2022008550

Set and designed in Garamond Premier Pro
by Mikala R. Kolander.

Contents

List of Illustrations. vii

Foreword. ix

Preface xi

Prologue 1

1. The Launch of *Apollo 14*. 3

2. The Roosa Family. 6

3. Growing Up in Claremore 9

4. Life after Claremore 12

5. Flight Training 14

6. The Barrett Family 17

7. Meeting My Mother 19

8. Early Family Life 23

9. Houston, Texas 29

10. Getting on a Flight. 35

11. Apollo Casualties. 40

12. Pastimes 43

13. *Apollo 11* 53

14. *Apollo 13* 55

15. *Apollo 14* 58

16. Moon Trees 65

17. The Country Western Tapes 66

18. Postflight. 68

19. Tales from the Road 79

20. Apollo Launches 84

21. Astronaut Downtime 85

22. Growing Up after *Apollo 14* 89

23. The Apollo Groupie Scene 108

24. *Apollo 17* 110

25. Reflections of an Apollo Command
    Module Pilot 115

26. The Last Flight of Apollo 118

27. Leaving NASA. 120

28. My Father's Passing 122

    Epilogue 130

Illustrations

Following page 64

1. Uncle Danny and Stuart at Yellowstone National Park

2. Unidentified individual and Stuart with his dog, Skippy

3. Stuart and unidentified individual

4. My parents' wedding

5. Out celebrating a traditional Japanese meal

6. Family portrait with the addition of Allen

7. Stuart in front of an F-104 Starfighter

8. My father in his major's uniform in 1967

9. The fifth selection of NASA astronauts, the "Original 19"

10. Jungle survival training

11. Training for splashdown recovery

12. Mitchell, Shepard, and Roosa in front of the *Apollo 14* Saturn V

13. My father's parents peer into the *Apollo 14* command module

14. Just a few hours from launch, my father looks so relaxed

15. The *Apollo 14* crew in a jovial mood

16. *Apollo 14* at liftoff

17. *Apollo 14* at liftoff

18. Launch day at the astronaut viewing area

19. Launch day photo op of the wives

20. My father flying solo around the moon

21. *Antares*, separated from the command module to begin its descent

22. My father looking at the checklist during his mission

23. My father getting ready to use the sexton

24. My father drinking orange juice

25. Splashdown

26. Back on Earth

27. Speaking at a press conference

28. My father receiving back the container used to carry the moon tree seeds

29. Stuart Roosa Day in his hometown of Claremore, Oklahoma

Foreword

Fifty years after the launch of *Apollo 14*, the audacity of the Apollo program still stands—and astounds—as a testament to the tenacity and ingenuity of humankind. Stu Roosa and I were among the privileged few who flew those space missions, and both of us made it to lunar orbit—in my case, twice. While only twenty-four men flew those Apollo flights and "slipped the surly bonds of Earth," their achievement was the result not only of thousands of scientists and engineers working steadfastly toward achieving an assassinated president's national goal but of the families that supported them.

We astronauts were family men, too, and our wives and children also "lived" the missions. Training for Apollo meant the astronauts were away from home for days, weeks, even months at a time. They also accepted that our passion for aviation and space meant they would live the highs of a successful flight—being invited to the White House and accompanying us on endless ticker tape parades in towns across America and the world—but also living with the possibility of a tragic and early death for the greater good of the nation. The Apollo kids often went to school with other astronaut kids—including those who had lost their fathers in training accidents that were a crucial part of the program. The moms, like Marilyn Lovell and Joan Roosa, ran the families during those long periods of time the fathers were away, as they did when we were military pilots. These absences and the background threat of never coming home were undoubtedly part of the learning experiences of the kids, but it was just the life we led. The families not only lived in the constant shadow of absent husbands and fathers; they did all of this on the meager wage of a military officer. It's high time their voices were heard.

Son of Apollo is a timely recounting of that extraordinary, unprecedented era of space exploration. It is told by Christopher Roosa, the eldest son of my good friend and fellow astronaut, Stu Roosa, command module pilot of *Apollo 14*.

He was an air force guy, and I was in the navy. But of course, Stu and I trained together on numerous occasions, and we shared the same ambitions and risks.

As Christopher remembers, Stu Roosa was originally scheduled to fly the ship that was used on my mission, *Apollo 13*, the mission that NASA subsequently called a "successful failure." As a result, Stu's intense training time in the simulator was instrumental in establishing the procedures that helped get us home.

Stu and Joan Roosa were our good friends. My wife, Marilyn, and I traveled the world both sightseeing and hunting with the Roosas, often with our families, and I came to know Christopher, the author of this book, through those adventures. In 1969 Stu and I took my son Jay and Christopher deer hunting for the first time.

It's time we heard the stories of children like Christopher who intimately experienced that defining era of human and technological achievement. No history can be understood through a singular voice. In *Son of Apollo*, Christopher Roosa relates the launch of *Apollo 14* and the events that led up to and after that historic moment. As children of the space program, Christopher and his younger siblings were able to witness from a unique vantage point such events as meeting presidents, captains of industry, and the popular stars of stage and screen who were attracted to the aura surrounding the Apollo astronauts. *Son of Apollo* provides an intimate account of what it was like to be a child of an Apollo astronaut, whether these are the recollections of a small boy battling school bullies in the sixties or a young man exploring the back country of Texas with his brothers. I am grateful Christopher Roosa has given us a view from the family perspective. It is long overdue.

Today, as we move forward with new technologies and a reenlivened desire to return to deep space, let us not forget the sacrifices of the men, women, and children who helped get us there.

Capt. James A. (Jim) Lovell, USN (Ret.)

Capt. Jim Lovell was the commander of Apollo 13 *along with CMP Jack Swigert and LMP Fred Haise. An explosion on the way to the moon threated the life of the crew, but the mission returned safely to Earth in an unprecedented tale of survival. Jim's book* Lost Moon *was used as the basis of the movie* Apollo 13. *He also flew on* Apollo 8, *the first mission to orbit the moon, in 1968. He had earlier flown on* Gemini 7 *and* 12, *making Lovell the first person to fly into space four times.*

Preface

While I was on deployment to Kabul, Afghanistan, this book started as a conversation at our compound bar, the Talibar. There, conversations can be pretty brutal. We were living in a war zone. We knew people were out to kill us. We lived through threats every day from rockets, mortars, and vehicle-borne improvised explosive devices (VBIED). Some explosions would rock the building. We routinely had attacks against us on our compound. We lost friends; we honored their sacrifice; we moved forward. This was a war.

I would wake up each morning and, while brushing my teeth, look into the mirror, knowing that someone had been up all night trying to figure out how to kill me. Of course, it was not necessarily the same person each day—we had lots of enemies who wanted to figure out a way to find our demise. They just had to get lucky once; you had to get lucky 100 percent of the time. This was just another day in the war zone. It was the way we lived.

At the end of one long day, I found myself sitting next to a fellow Marine, call sign Skyline, sharing a few stories and drinks at the Talibar. Someone had asked for a story about the Apollo space program, and I answered it. Skyline looked at me and said, "I need to tell you something. My kids are grown, and we are about to be empty nesters. Your kids are so young that if you get killed over here, they will not *even* know who you were. They will have no memories of you. Just some pictures of you. Much less will they know these stories of your experiences growing up. You need to write them down for posterity." It was harsh, but it was true. At that time, my oldest twins were six; our middle child was two; and the youngest twins were eight months old. They were all too young to remember any interaction with me. I took his words to heart.

A fellow colleague, call sign Bit, also implored me to write down the stories and offered to help edit the first rough drafts. She was great and offered many useful suggestions. I also need to thank Moonstruck for all her support.

I also wish to thank my wonderful wife, Danielle, who, through her love, efforts, time, and support, made this book possible.

This book is dedicated to my five wonderful children—Barrett, Reese, Flynn, Isabella, and Ryan—so that they may know their daddy. At least now my kids will know the stories.

To all, I give my love.

Son of Apollo

Prologue

President Kennedy stood before Congress on 25 May 1961 and proposed that the United States "should commit itself to achieving the goal, before this decade is out, of landing a man on the moon and returning him safely to the earth."

At that time, the United States had a total of fifteen minutes and twenty-two seconds of total space time. The president gave the speech twenty days after Alan Shepard's *Mercury 7* flight. It was an incredibly bold statement and launched the United States and the Soviet Union into a space race.

On 12 September 1962 the president at Rice University said, "We choose to go to the moon in this decade and do the other things, not because they are easy, but because they are hard." It would turn out to be hard on the astronauts and their families.

At the peak of the Apollo program, there were only twenty-eight astronauts involved. Almost all were married with kids. The kids ranged in ages from babies to kids in their twenties. Each had unique experiences, but I am sure we also shared some of the same experiences. Our fathers were part of the greatest endeavor of exploration ever undertaken by humankind. Everyone knew the goal; you could walk outside at night and look up at it: the moon.

Like many American children, we were taught that success meant doing better than previous generations. Be the first in your family to graduate from high school or the first in your family to go to college; but how do you surpass the legacy of a father who went to the moon? It was a big shadow to step out from.

Some astronauts' kids became involved in drugs, some had brushes with the law, some battled with depression, and some committed suicide. Others did well in business, the military, or other professions. While outsiders might believe that being an astronaut's kid would be living a fairytale, it was not. Sure, we had many unique experiences, but there was a price to be paid. In

1972 my father was home for six days out of the year. Most of us grew up with our fathers away training for the Apollo program. Life was not a fairytale.

This is my story of our family's journey through one of our nation's greatest times, the Apollo space program. While I share the experience from my perspective, my parents shared intimate details on some of their fellow astronauts and friends, to include details of divorces, infidelities, and other personal facts. Many of those details are associated with astronauts who are still living. I intentionally do not reveal names out of respect for their privacy.

In our family, we referred to our parents as Daddy and Mama. They both hated the names Dad and Mom. For the purpose of this story, I will refer to them as my father and my mother, which is how I refer to them when I talk to my friends about them. They both truly loved each other. While they had their ups and downs, as with any married couple, they served as an outstanding example to us all.

At the time, the astronauts were called our all-American heroes. So we used to joke that the kids must be the all-American kids. This book is the experiences and memories of just one of those kids who grew up in the Apollo era; thus, I am a son of Apollo.

Let the journey begin.

1

The Launch of *Apollo 14*

"In thirty seconds, we will resume the count," Cape Kennedy mission launch control announced.

Apollo 14 had been on a weather hold for forty minutes and two seconds. The original launch time was 3:23 p.m., 31 January 1971. The mission had a launch window of just less than four hours. If it missed the launch window, the mission would be delayed until March. NASA planned to land at the Fra Mauro Highlands. *Apollo 11* and *Apollo 12* had landed on flat areas of the moon, so NASA wanted to collect lunar samples from the more mountainous areas of the moon.

Fra Mauro was originally the intended landing site of *Apollo 13*. With *Apollo 13* having to abort its landing due to explosions of the oxygen tanks in the service module, the landing site dictated much of the launch windows. The shadows on the lunar surface had to be perfect to help the crew sight the large boulders at their landing site.

Apollo 12 had launched during poor weather, and the Saturn V rocket was twice struck by lightning. This caused the instrumentation and navigational systems to be knocked off-line. After these incidents, NASA tightened the launch rules for poor weather. As a result, *Apollo 14* was the first Apollo mission to experience a weather hold.

The Saturn V is hard to describe to someone who did not experience a launch. I say "experience" rather than "see" because it is a multisensory event. The Saturn V was imposing—some thirty-six stories tall. That in itself overwhelms. But it was also a living, breathing machine. The white puffs of venting liquid oxygen give it life, making it seem something like a wild beast ready to take flight. As the count continues, the venting stops, to allow the pressure to build in the oxygen tanks. Then you know you are getting closer to liftoff.

I stood with my family, in the astronaut viewing area at Cape Kennedy to watch my father launch into space. I was eleven years old. Three miles from the launchpad was the closest point that spectators were allowed to be positioned. I had been here once before for the launch of *Apollo 11*. At that time, I didn't know that he had already been selected for a prime crew. Originally, his crew was assigned to fly on *Apollo 13*, but NASA management changed the assignment and swapped the prime crews of *Apollo 13* and *Apollo 14*.

As I looked at this Saturn V rocket, I knew that Major Stuart Roosa, U.S. Air Force, was sitting on top of it, as the command module pilot, along with Captain Alan Shepard, U.S. Navy, as the mission commander and Commander Ed Mitchell, U.S. Navy, as the lunar module pilot. Alan Shepard, our first American in space, was head of the Astronaut Office. As the commander, Shepard personally selected his crew, and out of all the astronauts in the office, he had chosen my father for the crew. It was the crew with the least amount of space experience. Alan Shepard had fifteen minutes of space time from his Mercury flight, and the other two had none. They were referred to as "the old man and the two rookies." Daddy would joke, "At least Shepard was smart enough to pick an air force guy to be able to figure out how to fly them to the moon and back."

With a red, white, and blue theme, the boys were all dressed up in nice blazers, and my mother and sister in dresses. I remember asking my mother what I was supposed to say when the rocket lifted off. She replied, "Godspeed." Daddy had a penchant for not running on time, so my mother joked, "Stu can't even get his rocket off on time. The only event he was ever on time for was our wedding."

My father would say, "You don't want a flight hold. Just too many other things can go wrong that might get you pulled off the crew. All you want is those hold down arms to release, so you can be on your way. There is no turning back at that point. You are committed." On the previous mission, Ken Mattingly was pulled from *Apollo 13* just three days before liftoff.

The mission launch control announcer continued, "We are coming up on T minus sixty seconds; the tanks are pressurized. Mark. T minus sixty seconds, we are still go for launch."

"T minus fifty seconds and counting; we have now gone on internal power."

"T minus forty seconds; Shepard reports putting the final guidance alignment. We are still go for launch."

"Thirty seconds and counting; Stu Roosa just said, 'Thanks, it has been a good count.'"

"Twenty-five seconds and counting; we are still a go."

"Twenty seconds, guidance alert; the guidance system is now going internal."

My father said that when the count restarted, all he could see were the dark clouds out the capsule window. But the rocket trajectory had them bypassing the clouds.

The countdown continued, "Ten, nine, eight, seven, ignition sequence started." The flames of the Saturn V first stage came to life. "Five, four, three, two, one, we have launch commit. We had liftoff at three minutes past the hour!"

The hold-down arms released, and the Saturn V slowly began to move. Huge frozen sheets of ice, created by the subzero temperatures of the liquid oxygen tanks, fell from the sides of the rocket.

At the first hint of the flames, my heart started pounding hard in my chest, and my hands shook from the adrenaline rush. I remember standing off by myself in the field of the astronaut viewing area and yelling and yelling at the top of my lungs, "Godspeed! Godspeed! Godspeed!" over and over again. As if my yelling was willing the rocket to begin to move. The brightness of the flames almost hurt your eyes as you watched it.

"The tower is clear." The first safety concern that the rocket might hit the tower during liftoff and damage the rocket was allayed.

"Houston is now controlling," Houston mission control announced.

Watching the launch gave you a firsthand experience of the difference between the speed of light and sound. You could see the flames of the launch immediately, but the sound of the engines was always delayed, since the sound traveled slower over the three miles to the viewing area.

Then the sound hit. The roar of the engine noise poured over my body; like a tuning fork, it rattled my body. The deafening noise of the five F-1 engines of the first stage drowned out my shouts. The ground shook; the popping noise of the engines thrust so loud in my ears; other people yelled, screamed, and clapped.

The rocket passed up through the clouds. The first stage burnt six million pounds of fuel in the first two minutes. I kept yelling. When the rocket went out of sight, I remember turning and seeing a kid about my age, eleven, looking at me. He was giving me a strange look. I looked back at him and wondered, "Does he think I am nuts for screaming so much, or is he looking at me, knowing my father is on top of that rocket?"

It didn't matter. My father was on that rocket, and I was now a son of Apollo.

2

The Roosa Family

The Roosa family came to the New World from Holland on a ship called the *Spotted Calf* in 1661. Albert Roosa and family settled in upper New York State. At some point, a Native American tribe is said to have kidnapped one of Albert's daughters. According to family lore, he organized a raiding party, and when they could not find her, Albert and his men reorganized and recruited a larger raiding party. They found the tribe and rescued his daughter. The Dutch governor of New York presented a declaration to Albert Roosa proclaiming him an "Intellectual Gentlemen" (whatever that meant). Much later, two Roosas fought in the Revolutionary War, one as an officer and the other as an enlisted man. Another Roosa fought in the War of 1812. Many Roosas fought in the Civil War, for the North. Military service by the Roosa family, from the Revolution to the modern day, is a family tradition.

Stuart Roosa was born into a family of very modest means. His father, Dewey Roosa, was one of ten kids in his family, living in a small mill town in upper New York State. Dewey was born on 24 November 1898. When he was four, he learned the news of the Wright brothers' first successful powered flight of an airplane. He described his childhood as one in which everyone fended for themselves, especially at dinnertime. In 1914, when he was sixteen, he decided to leave home. He would jokingly state, "With so many kids, it probably took them a week to figure out that I had left." He bicycled west, did odd jobs in small towns, and moved on. He was in Ohio when his bike broke down, and he decided to join the army. World War I had just ended. Dewey was sent to Germany to be part of the occupational forces. He learned engineering and how to survey the land. He became part of the U.S. Army Corps of Engineers. He was later sent to Panama to survey the area around the Panama Canal. He would tell us the story of how, deep in the jungle, one of his team members accidently shot himself in the leg with his .45 caliber pistol. Dewey cleaned up the wound, put a splint on the man's leg, and walked him miles back to camp. The corpsman was very impressed with Dewey's work. My grandfather was always happy to share that story. In the late 1920s he conducted the site survey of the Oologah Dam and reservoir, not far from Claremore, Oklahoma.

Lorine DeLozier was a nineteen-year-old, petite woman working in her family's department store in Oologah. Born 3 June 1909, she was almost ten years Dewey Roosa's junior and married him on 27 March 1929. She left Oklahoma for the first time as she followed Dewey on his surveying assignments.

Along the way, Daniel Dewey Roosa was born on 28 September 1931. Two years later, on 16 August 1933, Stuart Allen Roosa came into this world. He was born at Mercy Medical Center in the picturesque railroad town of Durango, Colorado. He was only born there through happenstance of Dewey being on a surveying assignment in Colorado.

Dewey surveyed from the Northwest United States and parts of Highway 1 all the way north to Alaska. Lorine would tell stories of living in lonely adobe huts and the hard and desolate areas they traveled. One time while Dewey was away, the hut they lived in collapsed. Lorine grabbed the two young children, and they made it out alive just in time.

In the midst of the Depression, the small Roosa family moved to Claremore, Oklahoma. Grandpa was lucky since he had retired from the Corps of Engineers and was receiving pension checks. The family purchased a small house and farm outside Claremore and raised chickens for their eggs. Grandma and Grandpa sold the eggs to local grocery stores. That part of Oklahoma is nicknamed Tornado Alley. My grandfather put his engineering skills to work and built the family a storm cellar. Made out of poured concrete, it was one of the very few that had a functioning drain. Most other storm cellars were just underground bunkers, often with several inches of rainwater inside. He used to say that they would run to the storm cellar and it would already be full of neighbors. Grandpa and his family would have to shove to get inside their own cellar.

Even with my grandfather's pension, they were considered poor. They had an outhouse, and the family would only bathe once a week on a Saturday night. My grandmother would heat up kettles of water and pour them into a big metal washtub. Everyone bathed in order of age. My father, as the young-est, was always last and bathed in the dirtiest water. One of Grandpa's favor-ite stories had to do with the chicken coop. One day when it rained, parts of the chicken coop flooded. A large group of rats were huddled in the only dry spot, in the corner of the coop. Daddy took his .22 caliber rifle and, with rat shot, killed them all; it was never good to have rats around the chickens. He called it "the day of the big rat kill."

3

Growing Up in Claremore

So many of my father's earliest experiences were small-town experiences. They are stories of tough times, close community, and family. But they were so formative of who he became as a man. I see in those stories now the seeds of the man he was to become—a man who wasn't afraid of a challenge, who could hold his own in the face of adversity, and who treasured his country and his family.

In many cases, the stories he told us reflected the physical and emotional harshness of his early life. For example, he and his older brother, Dan, shared a horse named Cyclone. The horse favored Dan and didn't much like my father—so much so that apparently the horse bucked him off one day and, while he was still in midair, reared up and kicked him for good measure. Of course, my father got up and persisted with riding him. Anything else in those days would have been unthinkable.

Even on the playground, kids played hard. He attended Justus Grade School, outside Claremore. His favorite activity at school was playing marbles on the playground. Daddy would show us how he would flick his thumb out of his fist to knock his opponent's marble out of the circle. Everyone played for keeps, and he would walk home with his pockets so full of marbles that his pants would start to fall down.

Whether it was related to these victories or not, one day his brother, Dan, got into a fight with a guy named Buck Johnson, the playground bully. While Buck was pummeling Dan, my father rushed to defend his brother, grabbing a cinder block and heaving the block into Buck's side. It didn't even slow Buck down. Dan lost that fight, but I do know it wasn't the last one he got into. Incidentally, we got to meet Buck in 1971 after my father's flight. The town of Claremore had a Stuart Roosa Day to honor him for his *Apollo 14* flight. Ironically, perhaps, Buck had become the sheriff of Claremore.

Other stories simply reflected the nature of small-town life. For instance, my father played on the Justus basketball team. They only had five players, so everyone played the entire game. There was one game when the whole team came down with diarrhea. He told us that a player would run off the court to use the restroom while the other four would do their best to hold the game until he could return. I asked him, "Did the refs call you for penalty, for just running right back onto the court?" He replied, "It was a small school; refs let you get away with a lot."

Like many small towns, there were those who never left and those who left but soon returned, finding the pace of big-town life too demanding. Highway 66 runs through Claremore. According to Daddy, his cousins would look at the road and say, "It ends in California, the Promised Land." Every few years, someone would go to California, and after a few years, they would return to Claremore. He was the only member of the family to move out and stay out, other than returning to visit his parents. My grandmother loved Oklahoma—"God's country," she would acclaim it. But my grandfather, who had traveled extensively with the Army Corps of Engineers, hated it. Sitting on their back porch, he would tell a young Stuart, "It's a big world, Son; get the hell out of Oklahoma." He didn't know then that his advice would lead my father to the moon.

People often ask me how my father came into flying. Was it something he'd always wanted to do? The answer is yes. He loved flying, and he pursued it with a single-minded determination from a young age. Recall that he was just eight years old when World War II started. During World War II, one day in downtown Claremore, he saw two air force pilots walking down the street in their leather bomber jackets. Stuart Roosa decided at that instant he wanted to be a fighter pilot.

My father loved to hunt, and it helped put food on the family table. He would talk about how hard he would have to work to purchase shotgun shells and how a friend, whose family had money, could just shoot thoughtlessly. My father developed into a great marksman.

At Claremore airport, there was an abandoned B-17 bomber, and Stuart Roosa would crawl into the cockpit and fly his imaginary bombing missions. He took to making balsa-wood airplane models and hung them with sewing thread from the ceiling of his bedroom.

During his junior year of high school, the family sold the farm and moved into the town of Claremore. It was the first time in his life that my father had indoor plumbing. He was very smart and made good grades. He was very thankful for a teacher named Mrs. Gassett, who taught geometry. Amazingly, he recalled a time when Mrs. Gassett had a bonus question on a test and only three students, including him, had the right answer.

These are the small victories of childhood, whether it was the Roosa brothers winning the Oklahoma State Woodworking Championship three years in a row (Dan for one year and Stuart for the next two years) or the time they "got up to no good" and managed to escape the wrath of the law.

On one occasion, during my father's senior year of high school, a group of friends decided they had had enough of the girls in town all being taken by the cadets attending the Oklahoma Military Academy, which was on the top of a hill in Claremore. A group of boys jumped into a convertible and raided the military academy's barracks, where it quickly turned into fistfights. But someone, somehow, turned on the fire hose, and they all managed to get away. The police received a tip from a drive-in waitress who described the kids in the car. The next day, with their parents present, they were called into the principal's office and arrested. Daddy talked about how his mother just kept "chipping," or nagging, at him. The boys were young, confused, and all ready to plead guilty.

One of the boys' fathers was the local photographer and did the school photographs and weddings. He told the boys to stop talking, that he was hiring a lawyer. My father really respected how this man took charge. They went to court, and my father told us how daunting it was to hear his name: "Roosa versus the Great State of Oklahoma." The lawyer did a great job, and they were found not guilty.

Years later my father went into the courthouse and looked up the court record. With a five-finger page rip, he handily removed any record of the case. He used to say to us, "If you get into trouble, call me first. We will get you taken care of, and then we will decide what type of punishment you will get. But I promise you, I will not 'chip' at you." I am not sure he ever forgave my grandmother for that incident.

It must have been one of those teenage experiences that gave him the following insight, though I never did get the details. He used to say, "If you are ever being chased and decide to plow into a cornfield to get away, make sure to keep your foot off the brakes so no one can see your brake lights." I knew that if I should ever find myself plowing into a cornfield, I would keep my foot off the brakes, but it never occurred to me to ask, "What were you doing that you decided to plow into a cornfield? Whom were you trying to get away from? So you got caught because your foot was on the brakes; what happened next?"

4

Life after Claremore

Stuart Roosa graduated from Claremore High in 1951 and went on to Oklahoma State University and the University of Arizona. He only needed two years of college to apply for the Air Force Aviation Cadet Program. He stayed in shape by swimming for one straight hour each day in the college pool. He had jobs digging holes to plant trees and digging up trees for replanting. One of the toughest jobs he ever talked about was working in a bowling alley as the pin racker.

His favorite job during college was working for the U.S. Forest Service as a firefighter. At the time, the first year as a firefighter was spent fighting fires on the ground. After that, one could apply to be a smoke jumper. Daddy would talk of jumping out of C-47 cargo airplanes in his fire gear. He said he would rather land in the trees than in the fields, because the fields always had hidden rocks. He would tell us how he would climb the big trees. To do this, he would put on climbing spurs and, using a big rope, work it up the tree, swinging with one arm, then taking a step, then swinging the rope with the other hand, and so on. He would reach a branch and have to transfer to a much smaller rope to get over it. It was nerve-racking. He had a great fear of heights.

After cutting the parachute out of the trees, he would bundle it up and put on a pack with all his firefighting and personal gear and hike toward the fire. They would put the fire out and then start walking toward the nearest road and the pickup point. Obviously, it was a job that required a high level of physical stamina, but it was also a job that required teamwork and emotional fortitude. The young men worked closely together and gave each other all their support.

At one point, for example, my father had bought a brand-new convertible. He was sitting at a traffic light in the right lane. Another convertible car pulled up alongside him, and the passenger flicked a cigarette butt onto the hood of his car. He got out and flicked it back into their car and drove on. He didn't realize he was being followed as he returned to the smoke jumpers' U.S. Forest Service base. Someone walked in and said, "Stu, there is a group of guys out here looking for you. What's up?" "They probably want to fight," he replied, telling the story about the cigarette butt. That was enough for the gang in the barracks. My father walked out, trailed by twenty lean, hard, muscle-bound smoke jumpers. The guys from the other car decided that discretion was the better part of valor and left.

At times, when we were out driving, our car would come to a stop at a train crossing. The crossing arms would come down, and we would watch as the long freight trains would cross in front of us. My father would always make the comment, "Look at those nice warm boxcars." He would regale us with the stories of his time working in the Northwest with the Forest Service and how he would ride the freight trains down to Arizona. He would tell us that if you could, you would want to catch a fruit train, especially a banana train. With their perishable produce, the fruit trains moved faster, having priority over cargo trains.

On one of his first attempts at catching a train, he met a hobo who gave him sage advice: "Always get off the train as it slows down into the rail yard and then walk around the yard and catch the train as it starts out of the yard and before picking up any speed." The old man said the yard workers took glee in beating up "hobo bums." My father said it was a mistake that he would have made. The old man also told him that you didn't want to ride in the boxcars, because if the train derailed, the boxcars would not offer much protection. One was better off traveling in an empty coal car—a long metal car with waist-high steel. Daddy would talk about how they caught an empty coal car and how in front of them was a car full of coal. All night long they had coal dust blowing on them as they hunkered down inside the car in the freezing breeze. "It was a miserable night," he recalled, and I could still see his mind drifting to that night as he told that story. From then on, he made sure only to catch empty boxcars, which may not have been safe but were certainly warmer. We may have been the only astronaut kids taught how to ride the rails.

5

Flight Training

After the 1953 firefighting season, Stuart Roosa walked into the U.S. Air Force recruitment office in a small town in Montana to apply for the Aviation Cadet Program. The old sergeant recruiter handed him the application. My father had had rheumatic fever as a child and listed it along with the many other illnesses he had had as kid. The recruiter looked over his shoulder, reached down, jerked the paper away, and tore it up. He handed my father a new application and said, "Son, there has never been anything wrong with you in your entire life." Thus began my father listing no to questions on his medical history on all medical forms. If his original application had been used, he never would have made it to NASA. The NASA doctors would have denied him from the start. Years later he was in the dentist chair at NASA, undergoing a routine checkup. The dentist looked at my father's teeth and then back at his health chart and then back at his teeth. Then the dentist looked at my father and said, "This is strange. You have these striations in your teeth that are indicative of someone having had rheumatic fever. However, your health record doesn't show you ever had it." Daddy looked him and said, "Hmmm … strange. Let's keep going."

In 1953 Stuart Roosa reported to Williams Air Force Base (AFB), Arizona, for pilot training in the Aviation Cadet Program. Daddy threw up the first time he flew in a T-6 trainer, but he knew he would get over suffering from airsickness. He graduated from Aviation Cadets and received his flight-training commission on 30 March 1955.

Lieutenant Roosa trained at both Del Rio AFB in Texas and Luke AFB in Arizona at the gunnery schools. It was common for the schools to use tracer rounds in training flights. Tracers are built with a small pyrotechnic charge in their casing. When fired, the powder ignited. The pyrotechnic composition burns very brightly, making the projectile trajectory visible to the naked eye during daylight and very bright during nighttime firing. On a belt-fed machine gun, tracers are set every fifth round on the belt. On one particular night mission, his aircraft had been loaded with belts containing all tracers. He talked about rolling in on the target and squeezing the trigger. Daddy was stunned at how many ricochets bounced back past and around his aircraft. He decided he wouldn't be any less aggressive; but he never loaded all tracer belts again. He didn't want to see how bad it was.

Another time, they were testing bunker busters. The idea was to come straight at the bunker, full throttle, and under six feet in altitude. Right before the target, the pilot would release the bomb and skip it into the bunker. If the pilot released the bomb too early, it could skip back up and hit the airplane. If the pilot released it too late, the bomb would hit the area before the bunker and not penetrate the bunker. Thankfully, this was a short-lived idea.

He told us of a time when he was a flight lead and conducting a low-level flight run. He would start out by saying, "We were low." His wingman, who is not supposed to be lower in altitude than the flight lead, called my father on the radio, "I think I just hit a tree." My father's first reaction was, "Either you hit a tree, or you didn't hit a tree!" He pulled up, gaining some altitude, and then did a wingover to inspect his wingman's craft. There was a large tree branch jammed into the wing. My father said, "The squadron executive officer ordered us both to come in over the weekend and wax and polish the squadron commander's aircraft. May not have been a legal order, but we did it, and that was the end of it."

My father told us of the power of prayer in saving his aircraft. Once, when he was coming back from a training mission, he was low on fuel. As he told it, he knew the engines would flame out at any second. He prayed to the Virgin Mary to help him get down safely. As he was landing the plane, the engines flamed out on the runway. The air force did an investigation and made the determination that the series of right-hand turns he had made prior to landing had "somehow" allowed the fuel to keep going to the engines. He discounted that report and said, "My prayers were answered. That is the only reason why I safely landed."

After gunnery school, Lieutenant Roosa was assigned to Langley AFB, Virginia. There he flew the Republic F-84 Thunderstreak and North American F-100 Super Sabre, training to deliver nuclear weapons against the Soviet Union. His target was in Poland. He told us, if given the order, his mission was to fly toward the target, pitch up and loft the nuclear bomb toward the target, and then get out of there as fast as he could to make the return trip as far as possible before running out of fuel. The air force expected pilots to bail out and then find their way back to friendly lines, and that was what he planned to do.

6

The Barrett Family

My mother's childhood was exactly the opposite of my father's. The Barrett family immigrated from Mallow, Ireland, in the 1840s and settled in northern Mississippi. The land around there became known as Barrett Ridge. My grandfather John Thomas Barrett was born in 1898 at Barrett Ridge. He fought in World War I, and after he came back, he studied veterinarian medicine. He was referred to as "Doc."

Mary Falls was born 18 June 1898, in Boonville, Mississippi. Her mother was one of the first women in Mississippi to graduate from college. She was known as "Mammy." Mammy would tell the story of seeing this "gorgeous" man walking into the post office. She told her mother then, "I am going to marry that man." And she did.

Rumor had it that the Falls side of the family could trace itself back to the times of the founding fathers. Supposedly, Daniel Boone married into the family. However, the other side of the family had less auspicious lineage—my uncle Jack Reese could trace his family back to Jesse James.

On 23 May 1934 my mother, Joan Carol Barrett, was born in Booneville, Mississippi. The family later moved to Tupelo, Mississippi. Her siblings were considerably older than my mother. John Thomas Jr., who went by JT, was thirteen years older, Gloria was twelve years older, and Pattie was ten years older. By the time my mother has memories of her childhood, she was essentially an only child. JT was studying to become a doctor and was made a corpsman in the army during World War II. JT used to say that he saw so much blood during World War II that he no longer wanted to be a doctor. So instead, he became a lawyer. Both Gloria and Pattie were engaged when their fiancées were drafted. Uncle Jack married Gloria and was a sniper in the U.S. Army, and Uncle Knox married Patti and served in a scout platoon for a tank battalion. Doc gave them both permission to get married before their husbands deployed to Europe.

My mother's father, my grandfather Doc, was a large-animal veterinarian. He mainly attended to cattle. He said that the only time he was ever nervous around an animal was when a traveling circus passed through Tupelo and he had to crawl into a cage with a fully grown sick gorilla. He had a rule of not charging widows for services, because the Bible says to take care of widows.

The family had a large house in downtown Tupelo. Being a doctor, the family received an exemption from the rationing taking place during World War II.

My mother and Elvis Presley were in the same class. She would talk about Elvis bringing his guitar to show-and-tell and playing songs for the class. In her mind, Elvis was from the wrong side of the tracks. My mother grew up in a very different social stratum, with debutant parties and high society; her parents even hired some young kids to come over and play with her.

My mother would tell of the time after World War II when Detroit went back to making cars instead of tanks. Her family was one of the first to purchase a new car. One night, while returning to the house, my mother opened a gate and drove through. When she got out to close the gate, she didn't set the emergency brake correctly. The car rolled backward through the gate, and the open driver's door hit a tree and was damaged. The door never worked right again, since the dealer didn't have any spare parts to fix it, as it was so new. Later, as a teenager, my mother survived a vehicle rollover when both she and her date (the driver) fell asleep on the way home. My mother was always of the belief that the collar of her heavy fur coat saved her any real injury.

After high school, she attended the Mississippi State College for Women and graduated with a history degree. She was working on a master's degree for teaching, at Ole Miss, when she and her friends decided to leave Mississippi and try to get teaching jobs somewhere else in the United States.

7

Meeting My Mother

It's strange how fate brings people together. My father and mother were always amused by the synchronicity that brought them together.

My mother and five of her friends decided that they wanted to leave Mississippi and explore. To do this, they put a map of the United States on the wall, and one of the girls came up with the idea to throw a dart at it. The idea was that they would travel to wherever the dart landed. Blindfolded, one of the girls stepped forward and threw the fateful dart. It landed on the edge of the Atlantic Ocean at Norfolk, Virginia. My mother would often joke that "she nearly missed the whole damn country." With their minds decided, they all applied for seventh-grade teaching jobs in Norfolk, Virginia and were accepted. One of them had a cousin who was assigned to Langley AFB as a pilot. Langley AFB is in Norfolk, Virginia. The girl's mother had told him to take the ladies out and make sure they met "proper gentlemen." The group was eating dinner at the Langley officers' club one Friday night.

That same night, Lieutenant Roosa was down in the stag bar with his fellow pilots. They were celebrating being the first squadron on the base to get the new F-84F fighter. He went upstairs and approached my mother, who was sitting at a table. His opening line was "Have you ever tasted champagne and beer?" She thought, "How disgusting." "No," she firmly replied. "You should try it," he replied. She did. "I am surprised," she would later say, "it actually tasted pretty good."

My mother and her friends rented a house together down on the beach. It didn't take long for the pilots to discover six teachers, all eligible for dating. The fighter pilots, bomber pilots, and cargo pilots were all trying to work their action. One night, my father and his roommates had a party, and my mother was there. She finished her drink and asked him what she should do with the glass. He jokingly replied, "We throw them into the fireplace." Not knowing it was a joke, my mother turned and hurtled the glass into the fireplace. After that loud crash, he said to himself, "I like this woman." In order to have enough money to take her on dates, he was always borrowing money from the barrack's loan sharks. He would tell us, "Twenty dollars cost you forty on payday."

After only one month of dating, my father proposed to my mother. Unfortunately for him, she turned him down. She already had a longtime boyfriend, Van, in northern Mississippi, and everyone in her family expected that she would marry him.

One Friday night, Lieutenant Roosa was scheduled for a flight-training mission. In those days, they didn't have radar to track movements. They set up waypoints and called in their position. He had set up his waypoints, and at thirty minutes into his flight, he was supposed to be two hundred miles from the base.

In actuality, he was just circling over the base, although he called in his waypoints on time. At the appropriate time, when his flight plan had him at two hundred miles away, he turned off his airplane wing lights, pulled the throttle to neutral, and began a descent toward the teachers' house on the beach. He stated that the F-86 had a lag time of about four to five seconds to go from neutral to full afterburner. He was getting lower and lower, and then he shoved the throttle to full. The afterburner kicked in right over the house, and everyone inside was rocked by the loud explosion of the engine. The house shook violently. Drink glasses flew, and people hit the floor. People were screaming; someone yelled, "That was Stu. Call the tower and report him." Apparently, someone did call the tower, and that person was told that Lieutenant Roosa was over two hundred miles away. He continued to fly around and call in his appropriate waypoints and eventually landed. Years later I asked him how close he was to the house; I figured it must have been something like fifty feet. Daddy, not one to exaggerate, looked at me coolly and said, "Eighteen inches." No wonder the house rocked. I replied, "You were plus or minus nothing from plowing into the house." He responded with a smile, "You've got to be good," referring to his piloting skills.

My parents continued to date, and at some point, Daddy just started to refer to my mother as his fiancée. She didn't want to correct him in public, and that is how they became engaged. Eventually, she had to break the bad news to Van.

My uncle Jack and aunt Gloria, along with their seven children, had a large house in Sessums, Mississippi, located not far from Starkville. They owned a dairy farm, and at one end of the property was the house where my mother's parents, Doc and Mammy, along with Mammy's brother, Clarence, lived. Sessums was a very small community with a single store and mostly gravel roads.

My mother was sitting with Van out on the front porch swing. Unbeknownst to my mother, some of my cousins were hiding under the porch, and her sister and mother were listening in from the window, as my mother told Van that she was engaged to Stuart. He jumped up, yelling, "I am going to kill myself." He hurdled over the hedges in a mighty leap and drove away. He didn't kill himself, and my mother was now truly engaged to my father.

At some point, my father flew down to Mississippi to meet my mother's parents. He had her describe the location of Sessums, and he plotted it on his flight map. Columbus AFB was about twenty miles from Sessums. He flew over and looked down and made his plan. He circled and descended. The way he told it, he came over the top of the store at about five feet in full afterburner, buzzing the big house, the dairy barn, and my grandparents' house. The folks sitting on the porch of the store jumped underneath and thought, "The Russians are coming!"

All my cousins and my grandparents came running out into the yard. For years, they told the story of how Daddy put on an amazing airshow right over their house. After twisting, turning, and buzzing the house, he got low on fuel and decided to fly to Columbus AFB. Air show over, my mother left to go pick him up.

On the way to Columbus AFB, my father had a panic attack. He suddenly realized that the family had a dairy farm, and he was concerned that the loud noise from the airshow might have caused the cows to stop giving milk. He was now afraid he had made a bad impression on my mother's family. My mother arrived and assured him that everything was okay.

My grandfather thought it was great. To my father, that was the most important thing. At some point, the cousins and my grandfather went quail hunting, and they were equally impressed by his shooting skills. This only endeared him more to my mother's family and, most importantly, to the man from whom he would have to get permission to marry my mother, Doc.

The folks at the store started referring to him as Mr. Lieutenant and continued to do so for years. Over the years, Daddy used to laugh, "I am an Apollo astronaut and an air force colonel, but around here in Sessums, I am still just a lieutenant."

My parents got married on 21 September 1957. They had a convertible MG. Someone, as a joke, put sardines in the air-conditioning vents. According to my parents, the smell of rotting fish never got out of the car. So began the Roosa family adventure.

One of the first adventures was a temporary duty travel to Europe. My parents had saved up $3,000 to furnish their first house, when Lieutenant Roosa received a six-month assignment to Europe. He flew over with his jet, aerial refueling along the way. My mother went over on the *Queen Mary*. The boat hit rough seas, and the waiters had to wet the tablecloths so that the plates would not slide off.

My father would say, "We spent all our money and had a wonderful time, with many happy memories." Whenever Daddy was telling the story, he would wave his arms about, motioning to all the furniture in the living room, saying, "And here I am with a house full of furniture; it was a great decision."

8

Early Family Life

My father was convinced that his U.S. Air Force career was always off track. It was only in hindsight that he would see that these potentially unraveling events had actually brought him closer to being an astronaut. These events were always serendipitous.

The first time this happened was when the air force decided he needed to go back to school to finish his college degree. He had completed the two years that were required to be commissioned in the air force and was disappointed when they ordered him back to school. He felt it would take away from time in the cockpit and set him back in his career relative to his cohorts.

Stuart Roosa was accepted into the University of Colorado at Boulder, under the sponsorship of the Air Force Institute of Technology (AFIT), where he earned his bachelor of science degree in aeronautical engineering (with honors) on 26 August 1960. My father was always upset that his straight-A record was broken when he, so infuriated by a professor's extreme left-leaning teaching, stood up in the middle of class and called the professor "a Communist." As punishment, the teacher gave him a C for the class.

On 29 June 1959 I came into the world—Christopher Allen Roosa. According to my mother, Daddy snuck in after hours to the hospital room so that he could hold me for the first time. She would describe how he lovingly ran his fingers over my forehead. He got busted by a nurse who kicked him out, but he was happy.

After my father graduated from college, he was assigned to Tachikawa AFB, Japan, located not far from Tokyo. My mother would tell the story of flying there; we had to make a fueling stop in Hawaii. He held me to the window of the airplane while we circled over the field. After several minutes of circling, we began our descent to land. As I did not look well, he handed me back to my mother, and I suddenly puked all down the front of her dress. She was not happy. She had to stay in the same dress throughout the refueling stop and then the flight on to Tokyo.

At Tachikawa my father was disappointed at being assigned as the chief of service engineering, basically the base engineering officer. He spent a lot of time having to go over base schematics as part of the job. Again, he was frustrated that he was not in the cockpit of a fighter jet. He believed his career was once again off track.

On 2 January 1961 my brother John Dewey Roosa arrived; he would be called Jack. Fourteen months later, on 12 March 1962, my brother Stuart Allen Roosa Jr. was born; he is referred to as Allen. My mother would say she never felt so rich as when she was in Japan, where she had two full-time Japanese nannies and a cook.

Winters can be cold in Japan. Our house had a furnace in the living room, and my brothers' cribs were often placed close to the furnace to keep them warm. One night, for some unexplained reason, Daddy decided to move the cribs away from the furnace. The furnace blew up later in the night, sending black soot and tar all over the living room. Who knows if by fate he prevented my brothers from being injured or killed?

As I was getting older, the nannies taught me Japanese songs and words. My father came home one day and saw me speaking Japanese to the nannies. "We have to get back to the States," he apparently declared. "My kid is speaking Japanese before he can even speak English!"

After two years in Japan, he received his next set of orders. We were moving to Harrisburg, Pennsylvania, where he was assigned to Olmstead AFB, at Middletown Air Depot; it was a major support installation for the U.S. Air Force. Following repairs to F-101 Voodoo fighters, he would test-fly them before they were returned to a squadron. Again, he was disappointed he was not assigned to a fighter squadron, but as he would say, "At least I was back in the cockpit." Shortly after our arrival, he played another key role in our nation's history.

In October 1962 the Soviet Union, or USSR, placed ballistic missiles in Cuba, and the United States was on the verge of war with the Soviet Union. It has become known as the Cuban Missile Crisis. Captain Roosa, along with another pilot, took turns flying to Key West, Florida, to retrieve the reconnaissance film taken by the U-2 airplanes flying over Cuba. He then would immediately fly to Washington DC for the film to be processed and distributed to President Kennedy and the intelligence community. The U.S. Air Force issued him a .38 pistol to carry if he had to divert from landing in Washington DC, due to bad weather, so that he could protect the film. He scornfully said, "Divert? We're not going to divert. The country is on the brink of war, and someone thinks we will divert due to bad weather? It doesn't matter how bad the weather is; I am going to deliver that film to DC."

Also, during this time, my mother was pregnant with our sister. My mother was from Mississippi, but we had family who fought on both sides of the Civil War (or the War between the States, depending on where you were from). When I was very young, I remember meeting my great-grandfather. He was one hundred years old. He told stories of his father during "The War." There was one particular story he told about how my great-great-grandfather had been a successful farmer in northern Mississippi about the time of the war. He built his house with a trapdoor and a tunnel that led out into the field so he could quickly escape.

According to the story, Yankee patrols had come to the house several times to capture him, but each time, he escaped out that trapdoor. However, one night during the winter, there had been several days of rain, and the trapdoor had swelled. When the Yankee patrol showed up, he was unable to open the trapdoor and was finally captured. In the freezing night, he was put on a horse with his hands tied behind his back. On their way back to the Union lines, Confederate soldiers ambushed them. In the ensuing pandemonium, he spurred the horse and escaped. At some point, the horse went into a creek, and my great-great-grandfather fell off the horse, with his hands still bound, and only just managed to pull himself from the creek before he drowned. He worked his way back to his farm and caught pneumonia and died before he could tell anyone where the family fortune was buried on the farm.

Interestingly, as the story goes, in the 1930s tenant farmers on that property "suddenly became rich" and disappeared. The family assumed the tenant farmers had found the buried treasure. With all this Southern history, my mother was not going to be the first in her family to have a baby born north of the Mason-Dixon Line. She moved the family to Sessums, and Daddy would drive down when he could. My father joked that he made the trip so many times that he became friends with the highway patrolman who kept writing him speeding tickets. So on 23 July 1963, in the hospital in Starkville, Mississippi, my sister, Rosemary Delozier Roosa, was born.

One time, the Thunderbirds, the U.S. Air Force's precision-demonstration flight squadron, with their new F-4 fighters, came to the base for an airshow. Captain Roosa was selected to fly and warm up the crowd. His F-101 didn't necessitate the long distances to turn that the F-4 required, so he was able to keep his jet right over the crowd the whole time. We later heard that the chief of the Thunderbirds said, "We will never follow that guy again."

Again, my father was disappointed that he had not been assigned to a fighter squadron. He saw this as a setback in his career. But as before, this "setback" turned out to be an advantage; having accumulated so many flying hours, he was assigned his dream posting: the Experimental Fighter Test Pilot School at Edwards AFB, California. Captain Roosa was selected for the Aerospace Research Pilot School (ARPS) as part of the class of '64 Charlie—fourteen of America's top fighter pilots all competing against each other.

On any military base, one of the first actions for newly arrived families was for the commanding officer's wife to host a welcome party for the newly arrived wives. This was also true at Edwards Air Force Base.

Chuck Yeager, the first man to break the sound barrier, was my father's commanding officer. According to my mother, the newly arrived wives were invited to the Yeager home for tea. My mother said all the wives showed up at mid-morning in cocktail dresses, with long white gloves up to their elbows. Yeager's wife was named Glennis, after whom he'd named his Bell X-1 aircraft. They all nervously sat in the living room in their Sunday best while Glennis brought out a tray of teacups, saucers, and a large pot of tea. Then Glennis announced, "Y'all pour yourself a cup; I don't drink the stuff anymore." She then walked over to the refrigerator and opened a beer. My mother instantly liked her.

Some of my earliest recollections are of living on Edwards Air Force Base. It is located in the Mojave Desert, a huge dry lake bed that could also serve as a natural runway if needed in an emergency. My father talked of doing these "zoom" runs in the F-104 Starfighter. The test pilot would fly a particular arch at a certain speed and altitude. They would fly so high that the engines would flame out; the plane continued to climb until it would finally nose over and begin accelerating back toward earth. On the way down, the pilot would manually restart the engine. If it didn't restart, the pilot would do an emergency landing on the lake bed.

Our time at Edwards AFB, from 1964 to 1966, was an exciting time. One day, Daddy took me to watch the X-15 land. The X-15 was nothing more than a manned rocket that was carried under a B-52 bomber. The B-52 would release the X-15, which, when fired, could reach a peak altitude of 354,200 feet (about sixty-seven miles). The X-15A-2 attained a speed of Mach 6.72 (4,534 mph). Both models landed on skids. I watched as it dead-sticked in for a landing.

At Edwards sometimes our parents took us to a nighttime viewing spot in the desert. It was the testing stand for the F-1 engine, which would be used in the Apollo first stage. We could see the flame and hear the roar of the engine. Little did any of us know that in a few years my father would be sitting on top of five F-1s.

Also while we were at Edwards, the SR-71 Blackbird was unveiled. The black, two-manned reconnaissance aircraft flew above Mach 3. Daddy would talk about flying chase on the SR-71. He would fly around the SR-71, inspecting it, and the final check was to fly behind the aircraft and look into the engines. He said it was neat because the fuel burned a bright green, compared to the red of other jet engines. This was because of the high altitudes in which the SR-71 flew.

Another neat airplane, the B-70 bomber, was designed for supersonic flight. It looked like the Concord, but it had six engines at the rear.

With Edwards AFB being so isolated, the officers' club was the main source of entertainment, and pilots needed little excuse to throw a party. My parents would tell the story of the B-70 pilots throwing a party at the "O Club" (the officers' club) to celebrate the fact that they had flown faster than Mach 2. Always a competitive lot, the SR-71 pilots showed up wearing "Mach 3+" pins.

On another night, apparently after closing down the O Club, the pilots thought it would be fun to go wake up their 64-C colleague and the only marine in the class, Joe Wuetz, and his wife, Shirley. They proceeded to bang on the door of their home to wake them. As is custom, they were all invited in and given both drink and food. After some time, some of the pilots got up to leave, but Joe was having none of it. With a rifle in one hand, he announced, "Keep up the party; no one leaves until the paperboy comes." Certainly, my parents didn't expect to be out until dawn. The moral of the story was, "Never go and wake up the Wuetzs."

When I turned five years old, my parents hosted a big party, and my mother made some type of spiked green punch. Apparently, it was much more potent than people expected, and one of the 64-C pilots passed out on our washing machine. He told my mother never to make that stuff again. Everyone was drunk and wiped out.

But there were also solemn moments at Edwards AFB. Flying experimental aircraft is a dangerous business, and there were any number of wakes held at the O Club for pilots who had crashed while pushing the envelope of their airplane. My parents told us, "Everyone would show up, toast him, and the next day it was back to flying."

By the time he left Edwards, my father had over 5,500 hours of flying time, including 5,000 hours in jet aircraft. He was always bothered that flight experience was measured in hours. He would say, "We should compare takeoffs and landings. I get there three times faster than a transport pilot."

At Edwards AFB, I recall one boy on our street who was six and fairly big compared to me. This boy bullied all the kids on the block. He would shove me down, and I would come home whimpering. Daddy would tell me, "You cannot wrestle. You are not big enough. You will always lose. You need to make a fist and hit him as hard as you can right square in the nose. You hit him hard, and you keep hitting him. You knock him down, and then you kick him." One day, the bully was shoving around a group of five-year-olds, and I found the courage to hit him hard right on the nose. He busted out crying and ran off. I was the hero of the block, and it was the last time he picked on any of us.

I began kindergarten on Edwards AFB. Supposedly, the best and brightest pilots were at Edwards along with their families. All the kids in kindergarten were given a test, and the top twenty-five were placed in a special education program for first grade. I was one of those selected. For first grade, we participated in a new learning program called phonics. It was a whole new way of teaching English. In place of the twenty-six letters of the English language were more than forty phonemes. Although I placed at the top of the class, it forever made conventional spelling a challenge for me.

Families at Edwards AFB would do group outings together. It wasn't unusual for families to go into Death Valley for recreation. Sometimes this would be simply for a group picnic, although the not infrequent sandstorms would cover our food with sand. On other occasions, my father would take us into the desert to hunt jackrabbits, or we'd go exploring there. I have vivid memories of my brother Allen, then about three years old, running around the desert in his Batman costume. Another time, my brothers and I were in the desert, and we came across a scorpion, which I summarily stomped to death.

9

Houston, Texas

While at Edwards, my parents decided to take us to Sequoia National Park. On the way, my mother read in the newspaper that NASA was going to have another selection of astronauts. My parents turned the car around and headed back home so that my father could prepare his application.

Fortunately, he was selected for an interview. The interview process was quite lengthy and also involved undergoing a battery of tests. After all, as most folks are well aware, the Apollo astronauts needed to be at the peak of physical and emotional fitness, given the inherent uncertainty of space travel. According to my father, one of those tests was to submerge his arm in a freezing ice chest for minutes on end. He would say that the doctors pushed each other to see who could come up with the most outrageous physical, intellectual, and psychological tests.

My father also told us about doing the Rorschach inkblot test. The pilots would share information with each other when they came out of the test. "See color and movement," they told him. Apparently, color and movement indicated something desirable about the subject's personality and emotional functioning. Armed with that information, he was able to regale the examiners with what they needed to hear. "It's the Champs Elysee in Paris. I see the Arc de Triomphe. French flags are waving around. People are eating at the cafés and walking along the street." One psychiatrist complimented my father, saying, "You guys are amazing. No wonder you are going to be astronauts. I don't see anything like that." He would tell us, "It is all just a big game."

The most important part of the process was the astronaut panel interview headed up by one of the original Mercury astronauts, Deke Slayton. While Deke didn't fly on Mercury, he became head of the Astronaut Office. My father returned to Edwards AFB depressed. He told my mother that when he went into the interview, Deke asked only one question and then walked out in the middle of the interview process. He was confused about what he might have said to cause Deke to suddenly "walk out like that." He felt like his shot was over.

A few days later Charlie Duke, my father's best friend and another 64-C classmate, received his acceptance letter to NASA at the office. Again, Daddy was depressed that he didn't receive a letter. He went into the house and told my mother that Charlie Duke received a letter, and only then did he learn from her that NASA had also sent a letter to our house instead of to the office.

My father opened his letter to learn he was one of the nineteen astronauts selected for the class of 1966. He later discovered that the operations officer at Edwards was a former roommate and a best friend of Deke Slayton. Unbeknownst to my father, the squadron operations officer had called Deke and said, "You're a damn fool if you don't select Stu." Apparently, Deke was happy with my father's answer and didn't need to waste any more time with the interview, and that is why he walked out.

So my father went on to Houston. Over the summer of 1966, my mother had the movers come and pack us up. She drove us off to Sessums, Mississippi, to visit our grandparents and cousins, while Daddy was house hunting in Houston. He purchased a house in the El Lago subdivision, where many of the astronaut families lived, including Neil Armstrong.

It was a brick home with four bedrooms and two bathrooms, located on a curve. My father paid $25,000 for that house. Our address was 506 Cedar Lane. We had a patch of woods behind the house about seventy-five yards deep and a half mile long that was not part of our subdivision but would give us many hours of outdoor play and adventure, whether it was imaginary war battles, chasing snakes, climbing trees, riding our dirt bike over the mud puddles, or simply running amok.

From our house to El Lago Elementary was two miles. Other kids got carpools, but I had to ride my bike, since my mother needed to stay home with my little brothers and sister. I can remember on cold winter mornings riding to school and how much colder it was in the shadow of a house than between the houses in a patch of sunshine. I would sprint in the shadows and glide in the sunlight.

I can remember one time when it was very cold and I was pedaling against a very strong headwind. A carpool of friends passed me, and they honked. I just put my head back down and kept pedaling. Suddenly, BAM, they had pulled over to pick up someone, and I, with my head down, plowed right into the back of their car and found myself sprawled over the trunk. Everyone got a good laugh out of that, except for me. Daddy would tell stories about how he had to ride a horse one mile to get to school. My mother would tell him to be quiet; his son was going twice that distance on a bike.

So I started second grade with Mrs. Curtner. I jumped right into my studies with the enthusiasm of what I had learned in first grade. Even so, something was terribly wrong. One day, the school called my mother to discuss my performance: "We think your son may be dyslexic."

"How can that be?" My mother responded.

"He draws upside-down letters. He makes symbols in his writing," they explained.

"He was the top of his class at Edwards. He is not dyslexic," she said. "It's the phonics he was learning in his advanced class." So much for being in the "gifted class" at Edwards AFB! I've since learned that the years between first and sixth grade are key years for English cognition. I wonder how many other "gifted" students from Edwards forever struggled with conventional spelling and grammar?

One thing I was not blessed with was a singing voice. I had this in common with my father. It is one of the curiosities of the astronaut experience that folks would assume that an astronaut could do most anything, even if it had nothing to do with the space experience. Numerous times, postflight, he would be asked to achieve all sorts of feats: sing at the Grand Ole Opry, ride a horse bareback at a full gallop, pilot a 747 passenger jet, take a trick shot with a rifle. Most times, Daddy would politely decline, as he did at the Grand Old Opry. It's important to know your strengths, and he knew he just didn't sing that well. He would, however, sing to us at bedtime, particularly two old tunes: "I Ride Old Paint" and "Mr. Froggie Went a-Courtin'." We loved to hear him sing.

Every so often when my father was on stage giving a speech, someone would ask him to sing. Depending on the crowd, he would tell them, "I will speak the line, and you sing it." He would say a line, and the crowd would sing a variation of "On Top of Old Smokey." His air force version went

On top of old Smokey
All covered with snow
I lost my jet pilot
From a flying too low
He said that he loved me
Would do me no harm
One day he went ape
And purchased the farm
He put on an airshow
He did it for me
At attitude zero
He clobbered a tree
With throttle wide open
He made his last pass
With a 100 percent on
He busted, he crashed

The crowds always loved the sing-along.

Like my father, I dreaded singing, particularly the once-a-week choir practice held at my school, which had been renamed Ed White Elementary.

The teacher would go down the rows, and each student had to stand and
sing the scale, "1, 3, 5, 3, 1." A student would stand, sing, and then sit back
down. When the teacher came to me, I would stand and attempt to sing; the
class would laugh; and she would make me do it again, and then again, and
again. The classroom would be busting out laughing. This would go on for
minutes, until she realized she needed to hurry up and finish the other students. It was humiliating, and I hated that class.

My memories around that time are sporadic, and I suppose it isn't sur-
prising that the moments I recall are either the highs of my boyhood, such as
watching new aircraft with my father, or the lows of being a small boy with
all the insecurities of youth.

My height was an issue until well after I left school, but any insecurity I
may have experienced as a routine part of childhood as the result of being a
small kid was compounded by being the son of an astronaut. Even so, I stood
up for myself and my brothers and always made a brave show of it.

For example, at recess, most of my classmates would go out and play touch football. Since I was one of the smallest boys in the school, I would always be the last chosen. To overcome this embarrassment, I would bribe one of the more gifted athletes, Danny, with my lunchtime Popsicle, to pick me for his team.

It was in fifth grade that my dreams of following in my father's footsteps would be crushed. I was having trouble seeing, and my mother took me to the optometrist. I had not inherited my father's eyesight with twenty-ten vision but, instead, inherited my mother's, and she wore contacts. I received my first set of glasses that year. At that point, I knew I could never become a fighter pilot, since the U.S. Air Force didn't allow pilots to wear glasses. Not only was I one of the smallest boys in my school; I was now four-eyed.

These were the days of the Cold War. We had to practice getting under our desks for a nuclear attack. We would go to the theater and watch black-and-white movies of how bad the "Commies" were and how great we were. We did the Pledge of Allegiance every morning with our right hand over our hearts before the start of school. After school, our days were spent riding our bikes, playing pickup football or basketball games, doing "Evel Knievel" in the street over a makeshift jump, or fighting.

Down the street from our house lived a family named Sims. They had three boys about the same ages as my brothers and I. The Sims boys would start by picking on my sister, and she would cry and go to my parents. My father would call us in and tell us, "When you pick on one Roosa, you pick on all of them. Boys, go out and handle it." Using the same wisdom that my father had taught me at five years old, I didn't push or shove or attempt to wrestle. If the Sims boys started to pick on and shove my sister, I would punch them squarely in the face. Game on.

I took on Billy Sims, and my brothers each took on their counterparts. Banged and bruised, we returned home victorious. This process went on for several years. Again, as the (small) son of an astronaut, I seemed forever to be fighting to defend my siblings or fighting to defend myself. There were many kids who thought they would impress a girlfriend by going over and beating up the small "astronaut kid." Most came back with a broken nose.

Most of the kids I went to school with had fathers who worked at NASA or at corporations associated with NASA. Like most schools, we would have a science fair once a year. You could quickly tell whose parents were scientists or engineers who assisted their kids with their science projects. These kids would have elaborate, professional, graphic presentations with names like "Cold Fusion in a Can," "Proving Einstein Right by Bending Light," and "Harnessing the Sun, Micro Solar Power." Of course, we were only in the fifth grade; they couldn't have done these projects alone. My parents never helped me. Daddy was always away training, and my mother was busy raising a family of four kids. As a result, I would have a simple poster board with seashells I'd picked up off the beach at Cape Kennedy, glued on and hand labeled. Needless to say, I never won.

I did get some help from my paternal grandfather, whom we called Grandpa, to build a car for the Pinewood Derby of the Cub Scouts. We carved and sanded a block of balsa wood together. We took it to the post office to get it weighed, to make sure it was within regulations. We used graphite on the nails to reduce drag on the wheels. We decorated it with some beautiful sparkly purple paint and used some decals off a car model. In the end, we were not fast enough to place, but we did win third place for best design. I was very proud of that car.

10

Getting on a Flight

At NASA my father again thought his career was off track. NASA had an Astronaut Safety Review Board, whose role was to assign astronauts to review various parts of the Saturn V design. My father was disgruntled that he was not assigned to either the command module (CM) or the lunar excursion module (LEM). He believed those astronauts would be named first to a crew. His assignment was the first stage of the Saturn V. The Saturn V got its name from the first stage, with its five F-1 engines that produced over seven and a half million pounds of thrust. The first stage burned six million pounds of fuel in the first two and a half minutes and then separated and burned up in Earth's atmosphere.

My father said that there were very few people who knew it but that the first stage had steel rods that attached to the bottom of the rocket and to the launchpad. This function of the rods, according to my father, was to prevent the rocket from moving off-center in case one of the hold-down arms was out of sync with the others at the moment of release. The steel rods would initially hold the rocket in place and then stretch like taffy, allowing for a smooth lift-off. It all happened in nanoseconds. I have asked NASA engineers and curators about the rods, but no one had heard of them. However, it makes sense.

Early in my father's tenure at NASA, he was flying back from Cape Kennedy to Ellington AFB with Al Shepard. Shepard was the flight lead, and my father was his wingman. They were flying through some terrible thunderstorms, and my father was working hard to stay right on his wing. Daddy would say, "The flight lead can make the flight difficult or easier for the wingman." After they landed, my father said to Shepard, "That was a little rough up there; you didn't make it easy." Shepard's response was curt: "Sounds like a personal problem." Over time, Daddy came to believe that his flying that day had impressed Shepard and that it was one of the reasons Al Shepard chose him for his Apollo crew.

Over the course of the summer of 1970, we returned to Cape Kennedy, and Daddy took all the kids into the Vertical Assembly Building, commonly called the VAB. It was the largest scientific building ever to be built. The colossal Saturn V rockets were assembled inside, and then the "crawler," a massive machine with huge tank treads, was used to move the rocket to the launchpad, three miles away. We walked into this enormous building where several Saturn V rockets were in various stages of assembly for the Apollo program.

We rode in an elevator to the top of a particular rocket, and my father told us that this was the one he was going to be riding on. He had each child touch the outside of the command module, and he told us that if he died, like those in the *Apollo 1* fire, that he died doing what he loved. It was a little heady for children ranging from ten years of age down to six years old to comprehend, but it was also reassuring, because we knew that our father was involved in exploration and that exploration is never without risk. This turned out to be the rocket that was used in the ill-fated *Apollo 13*.

One time down at the Cape, we were driving around Pad 39-A, when my father pointed out the "slide for life." If the crew believed the rocket was going to explode, they could jump into this bucket and take a ride down a cable to a safe bunker area. I can remember one time sitting at our kitchen table and Daddy walking in, in his flight suit, returning from the Cape. He kissed my mother, who was cooking over the stove, and said, "We got the 'slide for life' human certified today."

My mother replied, "What does that mean?"

My father said, "The problem we had was the test dummies kept getting decapitated at the break cable. So I jumped into it and just did it."

My mother responded, "Well, I am glad you did not get decapitated."

As my father turned away to head for a shower, Daddy replied, "They couldn't get low enough. But now it is human certified, and we don't need to deal with it anymore." My father took risks. Everything about going to the moon was risky.

Months later we went back to the VAB and touched the command module of the *Apollo 14* rocket. He reminded us again that he was doing what he loved. We did not spend a lot of time at NASA with my father, but a few special visits, such as the visit to the command module in the VAB, stand out.

I remember another time he invited me to the flight simulator building. It was the summer of 1970 and a typical, hot Houston day. My mother told me that Daddy wanted me to come out to NASA, so she drove me in our big green nine-passenger Chevy station wagon, the ten-minute drive from our house to the NASA compound. I knew the way to my father's office building, but instead of going there, she pulled up outside a large warehouse-looking building. I got out of the car, and all alone, I walked up to the double doors and opened them. It was a large building, and inside, it was cold from the noisy air-conditioners running to cool the building.

Daddy walked up and said, "I thought we would go for a flight today," putting his arm around my shoulder and motioning toward a huge machine. I stared at this brown, oddly shaped monster of a contraption with a stairway leading up to the middle of it. It was about three stories tall and forty feet wide. In the middle of all that mess of machinery was a hatch. My father led me up the stairs, and as we stood on the platform, he opened the hatch. We crawled into the belly of this beast, which turned out to be a mock-up of the command module. The cool air was to prevent the computers and equipment from overheating. This command module simulator was a place where he had spent thousands of hours training.

My father let me go in first and had me move over to lie down in the left seat. He crawled in behind and lay down in the center seat. A NASA technician came up behind us and closed and locked the hatch. I looked up at the hundreds of switches and buttons on the flight panel. Daddy gave me a headset, and I overheard him talking to the controllers.

The inside of the command module is about the same size as an SUV. He showed me the hand controller: if I pushed it forward, the stars in my viewer would move upward and out of my view, as if I were descending; if I twisted it, my view moved to the right and left; if I pulled back, I would rise. Since there is no up or down in space, the command module would navigate by triangulating the stars.

My father had me attempt to dock the command module several times with the LEM. Each time, I failed. He would talk to the controller, and magically we would be set up again for another attempt. I would move the hand controller around and push the toggle for forward thrust. I would get frantic as I was closing in on the LEM, always faster than I wanted. One time, I put the right retro-rocket of the service module right through the LEM window. Of course, it was a simulator, so in reality, no damage was done.

As I have gotten older, I realize what was happening that day. My father knew how close the world came to losing the crew of *Apollo 13*. At particular moments, Daddy would tell me, "If I die doing this, I want you to know this is what I love doing." It was hard to understand those types of comments at eleven, but I would let him know that his words were registering. At that time, I didn't comprehend it, but I put the words to memory. Nowadays, I understand that my father was preparing me as the oldest son so that if something tragic happened on his mission, I was to comfort my brothers and sister with the knowledge that he was happy and thankful to be at such a point in history and to be part of it.

Back in the simulator, Daddy said, "I know you can shoot; let's see how well you can shoot a star." We moved to the far wall of the simulator, toward the sextant, which had an eyepiece that blocked out lights inside the command module from seeping into your vision. I placed my eye to the sextant and looked at the reticle. I watched as the stars moved quickly across the view. Some were bright; some were dim as I focused on the speed of the movement through my crosshairs. There were two toggles that moved the crosshairs up and down and right and left. There was a button to press on top of the right toggle, to mark the star position when you had the star centered.

One of the most important stars used by *Apollo 14* for navigation was the star Antares. It is the brightest star in the tail of the constellation Scorpio. It has two "guard" stars on either side of it. Often, in the nighttime sky, Daddy would point it out to me.

It was actually so important to the mission that Ed Mitchell, the *Apollo 14* lunar module pilot, named the LEM *Antares*. My father told me, "Okay, Antares is coming up. When it comes through, move the crosshairs onto it and push the button." I watched as the first guard star came across the view. Next, I knew the bright star was going to be my target. It came into view from my top right, moving toward the lower left. The stars were moving, the crosshairs were moving, and then at the precise moment, BANG, I pushed a button.

I pulled my head out of the sextant, and Daddy said, "Okay, let's see how close you got. The computer will let us know how far off you were." He pushed some buttons on a panel that looked like a calculator, and the numbers appeared: "00000." My father laughed and said, "Do you know how many thousands of hours I spent trying to make that shot? It was a hell of a shot. You nailed it." From Earth to Antares is roughly 550 light-years. Since 1 light-year is the distance that light travels over that time frame—approximately 6 trillion miles—

the total length of the shot was 3,300 trillion miles, or 3,300,000,000,000,000 miles. That was the one and only time I was ever inside the command module simulator with my father.

Daddy and I always shared a love of hunting. But as I think back over my life and recall that short special time inside the command module simulator alone with my father, I can still say, "That was the best shot I ever made." Range: 550 light-years!

11

Apollo Casualties

I graduated from Ed White Elementary and moved to Seabrook Intermediate School to begin the sixth grade. When we first moved from Edwards
AFB to our new subdivision in El Lago, Texas, the elementary was named El
Lago Elementary. Many astronaut families lived in El Lago. While we lived
there, the elementary school was renamed after Ed White, the first American to walk in space, who along with Gus Grissom and Roger Chaffee died
in the tragic *Apollo 1* fire. The White family lived not too far from the school.

My father was in the blockhouse at the launchpad, talking to the crew,
when the fire broke out. He was serving as "Stoney," the CAPCOM. Everyone
was having trouble communicating in the plugs-out test on 27 January 1967.
The test was to simulate the command module being on internal power. The
rocket was a Saturn 1B, located on Pad 34B. The mission was to stay in Earth's
orbit and test out the command module.

Suddenly, one of them was yelling into the headset, "We are burning up!"
My father's first reaction was to run out and look to make sure the elevator
was up top, at the crew gangway. His next reaction was thinking, "This guy is
going to hate hearing that tape in the morning." Fighter pilots always sound
cool on the radio, and there was clearly panic in his voice.

He sometimes lamented that NASA did a disservice to the *Apollo 1* families by implying that the three astronauts burned alive. He said, "Their suits
were white." While the accident investigation did show charring of the space
suits, the crew died of asphyxiation.

The few times we discussed it in the family, he said he was haunted by the
idea that he had that thought. He would say, "I had a man dying in my ear,
and my thoughts were how he would react when he heard what he'd said
over the mic. We just never expected a tragedy like that." He would sometimes softly say, "There are only two people who know what happened that

day, Deke Slayton and me, and I am not going to talk about it." Daddy had nightmares about it.

So every day, when I rode my bike to school, I was reminded of the risk of being an astronaut as I looked up in the parking lot at the name of the school: Ed White Elementary. Still, as a kid, even an astronaut kid, you don't dwell on it; you just keep riding your bike.

On 31 October 1964 Theodore Freeman was killed in the crash of a T-38 at Ellington AFB near Houston. The aircraft crashed after it was struck by a snow goose. The impact caused pieces of the canopy to enter both engines. He ejected but was too low for his parachute to deploy completely. His wife, Faith, survived him. My mother told the story that Faith Freeman first heard of her husband's death when a *Houston Chronicle* reporter came to her house and asked, "How does it feel to be the first astronaut widow?" In actuality, there was confusion between Deke Slayton, the *Houston Chronicle* management, and the reporter, on the timing of the death notification, which resulted in the reporter knocking on the door before Deke could tell her the bad news. The reporter was surprised to find out that Deke was not present, and as a result, Faith realized something terrible must have happened.

This episode embarrassed NASA, which decided they needed a better protocol going forward for handling the deaths of astronauts. For a start, they made sure that families would be informed quickly of the death and that other astronauts, not simply NASA staff, would inform them. My father had to perform this duty at least once.

On 6 June 1967 Ed Givens Jr. was driving his Volkswagen Beetle home from a meeting of the Quiet Birdmen fraternal organization, with two other officers, when he missed a sharp, unmarked turn and crashed into a ditch in Pearland, Texas, near the space center.

Givens died on the way to hospital. Daddy got a call from NASA telling him to inform Mr. Givens's wife, Ada, that Ed had died. The Givens lived in our same El Lago neighborhood. NASA wanted her to know before the press arrived, and our house was closest to the Givens'.

Because he was driving a VW Bug, my mother would often say, "If he had had an engine in the front, he might have lived. We will never buy a car that doesn't have an engine in the front." True to her word, we never did.

Of course, my father was also much involved in the events surrounding the *Apollo 13* flight. This is because the *Apollo 14* crew had originally been assigned to the *Apollo 13* flight. Due to a sequence of events by NASA management, they were reassigned to the prime crew of *Apollo 14*, with my father as command module pilot. The primary rumor was that NASA management wanted more time for Alan Shepard to train, due to his recent return to flight status.

On 13 April 1970 the world heard for the first time the immortal words, "Houston, we've had a problem here." *Apollo 13* had just ruptured an oxygen tank. The world held its breath while NASA worked to bring the crew home alive. My father's first thought when he heard from mission control about the situation was, "They're dead." Daddy used to say that if *Apollo 13* didn't have the LEM as a lifeboat, they never would have survived. For the next several days, he lived at mission control, helping to work out the solution to bring home his fellow astronauts. Other than a few folks at NASA, most people didn't know how truly precarious the fate of the crew was.

12

Pastimes

There is a perception of the Apollo astronauts as men of inordinate physical prowess, and this was certainly true of my father in many respects. While he had a singular vision for his passion of aviation, he also loved and exceled at a variety of outdoor and physical pursuits.

Daddy introduced me to many of the pastimes I treasure today. He was a self-made man and an outdoorsman, and he passed on many of those passions to his children. One of those was scuba diving.

When I was ten years old, our family was invited to go to the Las Brisas Hotel, in Acapulco, Mexico. It was a five-star hotel and had cabanas built on the hillside. Each cabana had its own swimming pool. They came in a variety of sizes and varying numbers of bedrooms. Each morning, workers came by and cleaned the pool and added flowers to float on the surface. The hotel had a lagoon built with a restaurant, a bar, and sunning areas with chairs. The lagoon had a small narrow entrance that reduced the waves coming from Acapulco Harbor. One could snorkel in the lagoon and watch fish swim back and forth.

They had instructors there who taught scuba diving lessons. I wanted to learn, so Daddy rented a tank. He had recently, as part of his NASA training, completed the navy scuba school. He rented a scuba tank and started by teaching me to "buddy breathe"—where two divers share one tank and pass the regulator back and forth. We did that for a while, until I was comfortable.

He then showed me how to empty my dive mask underwater. Even though my mask was full of water, he taught me to push the palm of my hand against the top of the mask and blow air out of my nose into the mask. The air would replace the water and make the mask usable underwater. He made me practice that numerous times. He strapped the weight belt on me so that I would sink. He had me sit on a rock in the lagoon about five feet below the surface. He let me breathe and get comfortable wearing all the gear. Next, he let me swim around in the shallow part of the lagoon. I did that for about fifteen to twenty minutes before he called me out. He helped me out of the water and took off my tank, mask, fins, and weight belt.

Finally, he threw it all into the water, about fifteen feet deep in the lagoon. "Dive down and put it all on," he instructed me. As I was instructed, I dove down and first put the regulator in my mouth so that I could start getting air. The salt water burned my eyes, but I looked around and found my mask. I put it on and cleared the water out. Now that I could see, I put on my fins and then the weight belt and finally maneuvered the scuba tank onto my back. I then swam over the edge and crawled out. My father was very proud.

He went over and talked to the head of the hotel scuba diving facility. His name was Carlos, and he had dived with the legendary scuba diver Jacques Cousteau. Carlos was a burly man. My father introduced us. The dive shop manager made me do the same routine. Daddy threw in the tank, mask, and other equipment. Down I dove and put it all on again. Carlos said, "Good enough."

The following day, Daddy took me down to the dock where Carlos and a group of scuba divers waited. I waved to my father, who was standing on the dock. We went out to do a dive on the wreck of a sunken ship in 110 feet of water. We got to the dive site and all jumped in. Carlos had me stay close to him.

We got down to the sunken ship. It was all rusty, and if you brushed up against any part of the ship, the rusted metal would cut your skin. You had to be careful with your legs. It was neat swimming in and out of the ship compartments. I don't know if it was because it was my first dive, but I really sucked down my oxygen. There is a valve you pull that allows for five more minutes of air from your tank. We were finishing up the dive when my tank went dry. I showed Carlos, and he nodded, took a deep breath, and handed me his regulator. He just calmly blew air bubbles out all the way to the surface. I was impressed. We got back on the boat and headed back to the dock. I found my family swimming in the lagoon. Thus, at ten, I did my first scuba dive to 110 feet, down to a sunken ship in Acapulco. My love for scuba diving continues.

My father was also an amazing hunter. As a child, he would come home from school; call his cocker spaniel, Skippy; and go wandering in the fields around their house. He loved to shoot rabbits and bobwhite quail. After the hunts, in the evening, he would sit by the radio and listen to the Grand Ole Opry while he picked the cockleburs from Skippy's fur.

It was my father who gave me my first BB gun, when I was seven years old. Then, as each of my three siblings came of age, they were also introduced to hunting.

When I was about seven years old, Daddy gave my mother a Browning 20-gauge shotgun for Christmas, along with hunting clothes, boots, and a shotgun vest. He told her it was time for her to learn to shoot. She went out and signed up for trapshooting lessons. He later gave her a Browning .234 rifle that became the kids' deer-hunting rifle.

When I was ten years old, we went deer hunting on a ranch in Texas. I was riding atop a truck that had been outfitted to hold a bench seat above the open cab. Those riding up top would lightly tap on the cab roof when they thought they saw something. The truck would stop, and everyone would look, some with binoculars, others through the scope of their rifles. The ranch was made up of open grassy fields surrounded by brush and trees. We would drive up and down the roads, looking for deer.

On this particular early morning, I was riding on top of the truck, holding my mother's Browning .243 rifle. There was a crossbar that we'd use to brace ourselves on the bench and that also functioned as a rifle rest. As we drove out of the woods and crossed into an open pasture, we spooked a herd of does in the field. The vehicle lurched to a stop.

My father was yelling, "Shoot, shoot, shoot." I lifted the rifle onto the crossbar and looked down through the scope as the deer raced across from my right to my left. I picked out the biggest one and cranked off a shot. My adrenaline was pumping. I saw a pile of dust rise in my scope. I raised my head.

Everyone in the truck was quiet, and Daddy enthusiastically raised his voice, "I don't believe it. What a shot! You hit it." There in the distance lay my first deer kill. My father was still stunned: "I need to pace this off." He got out and started walking, the vehicle moving along beside him. My heart was pounding, and I remember thinking, "Let's get to it." As we pulled up next to the deer, I could hear my father counting, "298, 299, 300."

When we reached the doe, he exclaimed, "Three hundred yards, on a flat-out dead run, and you hit it the neck." He reached out to shake my hand and said, "You are one hell of a shot." Over the years, my father would see me make longer shots, but he was always most impressed by that shot I made as a ten-year-old.

I remember another time, when we were hunting mule deer in the Big

Bend area of Texas. I was watching from a distance as Daddy and my brother Jack stalked a buck. I saw the buck take off running away from my father at breakneck speed, and my father spun around, raised his rifle, and with one shot nailed it.

Our father would also take us goose hunting. Those winter mornings, it was completely dark out when we would arrive at the frozen rice paddies. Tromping out into the middle of the field, we placed white cloth diapers, seemingly hundreds of them, around in the field as snow geese decoys. Just as the sun was coming up, we would crawl under a white sheet and wait as the caller would call in the geese.

We would yell, "IKE, IKE," making the sounds of the snow geese. I would try to find a small puddle of water and lie next to it. I would see the reflection of the geese as they were being lured closer and closer. The caller would say in a soft tone, "Here they come, here they come . . . Now!" And we would throw off the sheet and quickly try to target a goose before we started shooting. At times, my hands and feet were so numb from the cold that I lost all feeling. As they warmed back up, it would feel like millions of needles being jabbed into my hands and feet.

My father loved driving his Chevy Blazer in the muddy rice fields. He would be on the accelerator, laughing, as we all got thrown around, with the truck sliding around in the muck.

At another time, we were guests of John King, one of the wealthiest men in America in the early seventies. We were staying at his King Ranch in Colorado. We loved breakfast, because the ranch had a private chef, and each morning, he printed a menu for that day's meals. For breakfast, we would order silver dollar pancakes. They were made about the size of silver dollars, and we would get a big stack of them. I remember that the ranch house also had an indoor bowling alley. One afternoon, my father and mother were out on a walk next to the flowing creek. Daddy saw a trout in a pool and reached down and snatched it with his bare hands. He brought it back to the house, and I asked him, "What did you use to catch it?" As he held up his hand and wiggled his fingers, "Five finger lure" was his reply. The chef cooked it up for dinner.

While on that trip to Colorado, we went back to the town of my father's birthplace, Durango. They had a Stuart Roosa Day and dedicated a park to him. We were invited to ride on the Narrow Gauge Railroad, from Durango to Silverton, in a private refurbished railcar. It felt like you were traveling first-class in the late 1800s.

As kids, however, we didn't join our parents on all their hunting adventures, but my mother would tell us all about their adventures when they returned. My favorite story was the one she told about my father's jaguar hunt in Mexico, a story she wrote down for posterity, which I have included in full below. After all, my father's story is also her story, and it is clear that without her constant support, he could never have gone to the moon. Her voice is a vital part of this Apollo story.

In 1972 there were two couples, the Roosas and another couple who were invited to participate in a 10-day jaguar hunt. While the hunting was to be under arduous conditions, the decision was made that the wives were to stay in luxury at a resort in Puerto Vallarta. After only two days, we were surprised to see our husbands return. We learned that the other man had developed a bit of "tourista" and decided to spend a couple of days back at the hotel. Stuart, with a quirky grin on his face, said, "Joan—ride back to the camp with me and you can return to the hotel in a couple of days when I come back to get him."

Stuart's invitation sounded enticing, so off we went to the jungle. Along the way, we stopped at a roadside stand and bought a watermelon. It was dark when we got to the camp; thus, I couldn't see much of the surrounding area. In camp, there was a lighted table with a cup holding spoons in it. Stuart cut the melon with his hunting knife and handed me a piece. Then Stuart went to put my overnight bag in the tent. Meanwhile, I picked up a spoon and started eating a piece. He returned to the table just as I was finishing my piece of watermelon. Stuart reached over, snatching my spoon, and started eating with it. Perplexed, I asked him why he did that. His reply was, "That's the cleanest spoon in the camp."

Before dawn the next morning, the professional hunter, with his men and dogs, left camp with Stuart. Stuart told me to sleep as long as I wanted and that they would return sometime close to noon. Not being a late sleeper, I got up fairly soon after daylight and came out of the tent to come face-to-face with the meanest looking man I had ever seen in my life. I felt like the word "E-V-I-L" was flashing in red neon letters under his forehead.

While standing there with a huge knife in his hand, wearing a completely filthy undershirt, I could not help but notice his long dirty fingernails. He bowed to me and said something that sounded like "the cook."

"Ongry?" he asked. Realizing he was asking if I was "hungry," I shook my head no, and motioned to my watch. "Later," I replied. I noticed that there was a lawn chair sitting nearby and a little fire that had a pot of beans boiling on it. Tortillas were thrown and lying on the bare dirt around the fire. "The cook," was scrambling some eggs, onions, and tomatoes in a little tin pan.

I sat down in the lawn chair, and he squatted down across from me cooking the contents of the pan. I wasn't afraid of the cook because I knew if anything happened to "Señor Luna's" wife, he would be in deep trouble with the outfitter. As I surveyed the camp, I deduced there was no hole to bury garbage. Furthermore, when one needed to go to the bathroom, one took a roll of toilet paper and went out to find a nearby bush. Just then a breeze came up, and toilet paper was blowing all over everywhere. Quickly, I saw firsthand why Stuart thought the spoon was the cleanest thing in camp.

Watching the cook chopping onions, I noticed how every now and again, he swiped the knife across the front of his filthy undershirt that covered his large belly. I shifted in my chair, and he jumped up, looking at me, and said, "Ongry?" again.

While I would sit still as long as possible, finally I would have to move, and he would shout, "Ongry?" I quickly learned that if I moved a muscle, his action was going to be repeated. I decided that I would have to be near starvation before I wanted to eat anything he fixed. This was becoming one of the longest mornings that I have ever spent. A little before noon, Stuart and the crew came in with a jaguar! I was almost as relieved to see him, as I was after his splashdown in the Pacific Ocean, on the return of *Apollo 14*.

Hearing the story, they had spotted the jaguar lounging in a shallow pool of water. Interestingly, the jaguar is one of the few cats that likes to get wet. As they closed in on the jaguar, the cat jumped and quickly killed two of the dogs and then ran up a small tree. As Stuart attempted to move in close enough to get a shot, the jaguar suddenly leaped at Stuart. Not even having time to aim and firing from the hip, Stuart shot and killed the jaguar. The outfitter and his crew were in total awe of Stuart. Not only for going to the moon, but for his coolness of that shot. "Señor Luna" had only inches between himself and the jumping jaguar.

Watching before lunch, I noticed one man take off dirty socks and wash them out in a bucket before everyone started to eat. For lunch, I decided the only thing I was hungry for were the beans that had been simmering all morning. I figured that with all the boiling, they would be the safest things I could eat. While we were all eating lunch, I learned the regular cook had gotten a toothache, and left to go home. The outfitter had left the man in camp with me on strict orders to take good care of "Señor Luna's" wife.

Right after lunch, someone threw out the water in the bucket that had contained the dirty socks, they then poured in more water and started washing the lunch dishes, without even rinsing out the bucket in-between. While such was camp life, I yearned for my resort room.

After they had skinned and salted the hide, Stuart said it was too late to start back to Puerto Vallarta and that we would leave the next day. The road was not a good place to be after dark. Funny thing, that evening, the only thing I was hungry for was some more beans.

Stuart took me back to the hotel the next morning and before he left to return to the hunting camp with the other man, I loaded them up with things like crackers, cookies, and other items that came in tightly wrapped packages.

We still have the jaguar skin in our house, and when I see it, I am reminded of my mother, my father, and their adventurous life.

Bob Perkins and a friend from Coos Bay, Oregon, read in the newspaper about astronauts conducting geology training near Bend, Oregon. They followed the astronaut training bus out to the training site and followed along behind the astronauts. At lunch, box lunches were provided to the astronauts. My father noticed that the two did not have anything to eat, so he grabbed two boxes and introduced himself and gave them lunch. It was to begin a lifelong friendship with the person we would call "Uncle Bob."

Over time, they invited Bob to join them at the Cape Kennedy complex. To get a visitor pass, one had to list title and occupation. They started listing him as the junior senator from Oregon. Daddy used to say, "Nobody knows who the junior senator is from any state, much less Oregon." So to the astronaut corps, he became Senator Bob. He was so close to the family that when the *Apollo 14* rocket rolled out of the VAB on the crawler for the three-mile trip to the launchpad, there were four men standing on top of it: Alan Shepard, Ed Mitchell, Stuart Roosa (the crew), and Senator Bob.

When I was ten years old, my parents dropped me off at the airport, and I flew to Portland, Oregon, to visit Uncle Bob. We toured Mt. Hood, Crater Lake, the obsidian flows at Bend, and then down to Coos Bay. He owned the Coos Bay Inn, a hotel and restaurant. I loved eating in his restaurant. They had miniature shrimp cocktails. I had never seen shrimp so small. Uncle Bob was an avid golfer. He made his backyard into a par-three chipping golf course. Uncle Bob took me out to a ranch that had a minibike. I jumped on it and opened the throttle and popped the clutch. The bike took off in a wheelie. I panicked, and with the throttle wide-open, I hit a tree. The bike climbed up the tree and finally toppled back over on top of me. It was the first of many motorcycle wrecks to come.

Uncle Bob was in his sixties when we first met him, but he was full of energy. He had been a vaudeville magician in the 1930s and '40s. He could do amazing card tricks. His specialty was an ESP trick he did with his wife, Alice. She would be blindfolded, and he did the trick the same way each time. You could hand him anything, a dollar bill, a shoe, a news article, and he would start out the same way each time: "I have something in my hand."

"A dollar bill," she would reply.

"Correct. What is the domination of the bill?"

"One dollar," she would reply.

"Correct. What is the serial number on the bill?"

"A 22961157 C." To the amazement of everyone, she was correct, every time. Daddy would try constantly to pick an object, trying to shake things up, but Uncle Bob and Alice would fool him every time.

Uncle Bob created a golf newsletter called the *Oregon Golf News*. He would get to go to all the major golf tournaments, such as the U.S. Open and the Masters, by applying for a media pass. He enjoyed going to the press conferences and meeting the players. He did it for years. I asked how many subscribers he had. He laughed and replied, "One—just me." It was a great way to get into the golf tournaments.

Clearly, my father was blessed with great hand-eye coordination. It was what made him a great pilot. Sometimes it had unintended consequences. In the summer of 1974 the family went to Oregon. Daddy loved the Pacific Northwest and would reminisce about his smoke jumper days and fighting forest fires. During one visit to Oregon, Uncle Bob had arranged for us to go on a several-day float trip down the Rogue River.

We were spread out over several wooden boats, with a guide who rowed the boat. If it was hot, you could just jump overboard and float along in the cool river and then climb back into the boat. For amusement, each guest was given a Wrist-Rocket slingshot, and each boat had a big bucket of marbles. You would sling marbles toward the seagulls sitting on rocks and logs along the river. The seagulls would just fly away. No one ever managed to hit a seagull; it was simply a way for the river guides to pass the time with guests.

One time, one of the guides called out to my father, "Stu, you are an astronaut—see if you can hit that seagull on that log." Everyone watched as Daddy picked up the slingshot and took aim at the seagull about forty yards away. Now the boat was moving along at the same time. My father released the shot in an arching approach and hit the seagull right in the head. It fell dead immediately.

All the guides starting rowing quickly to leave the scene of the crime. Apparently, it was illegal to kill seagulls, but of course, no one had ever come close to hitting a seagull, let alone killing it. The lead driver said he'd been guiding tours on the river for twenty years and never had anyone hit a seagull before. My father, unintentionally, became a legend on the river that day.

There was another time, when we were all at some picnic grounds located right next to the edge of the Grand Canyon. In the picnic grounds were these chipmunks that ran amok through the picnickers. Everyone was trying to shoo them away. Many of them were running in and grabbing food from the picnic blankets. Daddy was walking back from the car when a chipmunk dashed across our picnic cloth. Daddy yelled at the chipmunk to clear out and casually kicked at it. With uncannily precise timing, his foot caught the chipmunk, and it flew about twenty feet over the edge of the cliff, unfortunately to its death. Many saw my father kick at it and were in awe and horror that the chipmunk went over the edge. From what I remember, the chipmunks left us alone after that.

In 1968 my father purchased a sixteen-foot Invader ski boat, with a big black outboard Mercury engine. My mother named it the *Just Right*. We would take it out on Taylor and Clear Lakes. We loved going out into Galveston Bay, because we would have to go under the big bridge at Kemah. The taller sailboats would blast their horn three times, and the bridge operator would shut down traffic and open the drawbridge. We thought that was the coolest thing.

I can remember one day, as I sat in my bedroom, Daddy was yelling, "Joan, Joan." I saw my father run by with a big package. I got up and followed him into the master bedroom, and my mother walked out of the bathroom suite and unwrapped the package. Daddy was excited, and my mother started laughing. For their wedding anniversary, my father had given her a shrimp net. She loved it; it was what she wanted. We dragged that shrimp net all over and never really caught any shrimp, but we would catch crabs.

Out in Galveston Bay is Red Fish Isle. It is mostly rocky and has few sandy beaches. The first time we went there, we anchored the boat in the shallow water on the ship-channel side of the island. Everyone was running around, collecting firewood, picking up oysters, wading back and forth from the boat, unloading picnic supplies. In the distance, a large container ship was going by on its way to Galveston Harbor. We were just watching the ship, not really thinking about anything, when suddenly the wake hit the shallow water and turned into waves. The waves started to lift the Invader and shove it to the shore. My father ran out and grabbed the stern railing and was bobbing up down in the waves while heaving with the boat to prevent it from being breach on the shore. In a few seconds, it was over. We never anchored on that side again.

My mother had decided we needed to get a bigger boat, because she wanted one with a head on board. She was tired of the ladies having to squat over the side. We got a yellow twenty-one-foot cabin cruiser that had a V-shaped cabin and a small head located at the top of the V. My mother named this boat the *Just Right II*. With this boat, we used to travel all the way to Galveston. We would overnight in the harbor. My parents and sister slept down in the cabin, and the boys slept on the deck part of the cabin roof. It was not very comfortable, and since we would do the trips in summer, it was hot and humid. We spent many hours out on the water, sometimes fishing, going to Red Fish Isle, or exploring the backwaters of Taylor Lake.

13

Apollo 11

I count it as my great fortune that I witnessed many Apollo launches. Of course, one of the most historic of these launches was that of *Apollo 11*, and I have vivid memories of that day.

While many families would line the road leading to Cape Kennedy or watch the launch from nearby Cocoa Beach, astronauts and their families viewed the launch from the astronaut viewing area.

The astronaut viewing area was the closest viewing area to the launchpad. It was three miles away, about as close as you can get to the launch and still be safe. The Saturn V rocket was the height of a thirty-six-story building. It weighed six and a half million pounds and generated seven and half million pounds of thrust at liftoff. Both liquid nitrogen and oxygen, the fuel for the rocket, made the outer aluminum skin of the rocket so cold that in the humid Florida air, the condensation on the outside of the rocket would turn to ice.

On the launch day for *Apollo 11*, the astronaut viewing area was filled with a big crowd, with guests of both the prime crew and the backup crew. I stood next to my father when I got to meet Charles Lindbergh, the first man to fly solo across the Atlantic. There were many VIPs there, and I wish my parents had taken photographs. But that wasn't their style. We later spent some time talking to the actor-comedian Jimmy Durante.

The announcer's voice over the loudspeakers went through the countdown: "Ten, nine . . . ignition sequence starts . . . six, five, four"—the rocket engines ignited, and they were the most intense light; the vibration of the launch shook off these huge hunks of ice from the rocket—"three, two, one, zero . . . LIFT-OFF! We have a liftoff, thirty-two minutes past the hour."

Apollo 11 slowly started to move upward in the sky. Then, all of a sudden, those watching the launch were hit with a massive wall of sound. The sheer energy of that sound jolted your adrenaline to kick in. My heart pounded and my body trembled. It was almost like being inside a huge popcorn maker. Pop! Pop! Pop! The ground shook like an earthquake. This wall of sound continued to pour over your body, and the rocket continued up and up, past the launch tower. Cheers rang out everywhere.

There were people yelling with excitement, clapping, hugging, standing in awe, as we watched the rocket jettison the first stage, just two minutes into the fight and six million pounds lighter. The intensity of the engine flames and the noise never came across on television, not like they did in real life. We followed the flight on TV. We watched Neil step on the moon. My father always said that Neil Armstrong was the perfect choice as the first person to step on the moon. Little did I know then that my father was scheduled to fly only two flights later, on *Apollo 13*.

14

Apollo 13

I can remember my mother calling me into the master bedroom when I was ten years old. She sat me down on my parents' bed. She told me that Daddy was going to fly on *Apollo 13*. She told me that I could not tell anyone, and I was going to keep that secret.

So the *Apollo 13* crew of Alan Shepard, as commander; Stuart Roosa, as command module pilot; and Ed Mitchell, as lunar module pilot, were all training in secrecy. The *Apollo 11* crew had been announced, along with the *Apollo 12* crew. NASA didn't want to take any attention away from the *Apollo 11* crew, with the news of Al Shepard, America's first astronaut, going back to space.

Given the secrecy surrounding the launch and my promise to my mother, I was horrified when a classmate confronted me in the school hallway: "Walter Cronkite announced last night on the news that your father is going to fly on *Apollo 13*." I stumbled for an answer and said that I didn't know what he was talking about. I rode my bike home terrified that my mother would think I had told someone. When I got home and told my mother, she told me not to worry; she knew that it wasn't me and that someone at NASA had leaked it.

This announcement forced senior NASA management to make a decision. They decided that since Al Shepard had only recently had surgery to ease his Ménière's disease, a condition in which fluid pressure builds up in the inner ear, he needed more time before he could go back on full flight status. Thus, the prime crews of *Apollo 13* and *Apollo 14* were swapped. Both Alan Shepard and my mother were furious at the announcement. My father just said, "Hey, I am still on a prime crew." After the crews were swapped out, my father continued to train for *Apollo 14*. Many hundreds, if not thousands, of hours were spent inside the simulator.

My father would tell the story about being picked for *Apollo 13*. The normal progression was to get assigned to a support crew for a mission; then about three missions later, to a backup crew; then three missions after that, to a prime crew. My father was on the support crew for *Apollo 9*, in March 1969. His motto was, "If the crew is awake, I am awake at mission control." The support crew served as the CAPCOM (the capsule communicator), or the liaison between an in-space crew and mission control.

One Saturday morning, my father was sitting at this desk in the Astronaut Office, and Alan Shepard walked into his office. Daddy used to say that Alan Shepard was the most intimidating man in the world. He could just look at you and put fear into you. Many of the astronauts would not pass him in the hallway. Instead, they would duck into a nearby office, whispering, "Shepard," so that everyone would nod their head and go back to work. Alan Shepard said to my father, "If you don't mind riding with an old man, I would like for you to be my command module pilot on *Apollo 13*." In my father's mind, he was sure that he had missed something, because he was focused on getting on a backup crew. He replied, "Did you say prime or backup crew?" Shepard looked at him and said with complete disgust, "I back up no man," and walked out of my father's office.

My father then realized he had been picked to fly on a prime crew and had screwed it up in the first ten seconds with his commander. My father went down to his red Chevy Blazer, got into the car, and screamed, "I'm going to fly!" while banging his hands on the dashboard in pure glee. He was the first Apollo astronaut to be selected for a prime crew without ever having served on a backup crew.

My father was at home when he got the telephone call that an explosion had occurred on *Apollo 13*. His first thought was, "The crew is dead." He rushed off to mission control. Since he had been training in secrecy for *Apollo 13*, he had more time in the command module simulator than anyone. He and the other astronauts had to work out the procedures to close down and restart the batteries for the mission home. Daddy talked about the many times he would just flip switches trying to eke out another volt. Ed Mitchell had to get into the LEM simulator to try to figure out how to use the LEM to fly with the command module back toward Earth. None of this had ever been simulated. None of this had ever been considered. This was NASA at "its finest hour."

While *Apollo 13* returned safely to Earth, there was an issue with my father, stemming from the fact that Ken Mattingly had been pulled as the command module pilot only three days before liftoff. Ken had been exposed to the measles by Charlie Duke, so Jack Swigert, Ken's backup, took the command module slot. Feeling sorry for Ken's situation, Buzz Aldrin started a movement to have Ken now fly as the command module pilot of *Apollo 14*. That put my father in a terrible position. "I didn't know what to do," he'd say. "Here is the second man to have walked on the moon, now advocating for me to be swapped out. I am a nobody. Never been in space." He said it finally reached a point that he thought he should address the issue directly with Alan Shepard, the head of the astronaut office and the commander of *Apollo 14*. Apparently, Alan Shepard was not aware of the movement and told him everything was fine. Then Shepard called Buzz to his office. While my father did know what Shepard said to Buzz, the gist of it was, "Don't screw with my crew." The movement died, and Ken later flew on *Apollo 16* beside Charlie Duke— the very man who had exposed Ken to the measles.

15

Apollo 14

After the explosion and the safe return of the *Apollo 13* crew, the focus and the pressure of the entire Apollo program then shifted to the *Apollo 14* mission. If the Apollo program was to continue, the *Apollo 14* crew would need to get safely to the moon, land, and return. They trained hard. My father would say, "If you had to place the space program on one man's shoulders, that man would be Alan Shepard."

A few months before my father's flight, he brought home several large boxes filled with freeze-dried space food. Each item was a vacuum-packed plastic bag with a little label stating the product. Some were meant to be eaten without rehydration, and others stated how many ounces of either warm or cold water were needed and how long to wait for the rehydration process. Those that had to be rehydrated had a short plastic adapter teat that would allow a water gun to insert either warm or cold water. The Apollo program was the first to have the option of warm water, heated by the spacecraft fuel cells.

The boxes contained items like bacon, ice cream, BBQ bites, pineapple fruit cake, brownies, mixed vegetables, chicken and rice, butterscotch pudding, graham cracker cubes, frozen fruit drink (Tang), and nutrition bars—none of which looked appealing. But they were neat to look at with all the various selections.

My parents invited over a group of friends, and Daddy explained the ground rules. My mother would provide the hot or cold water to reconstitute the meal, and then after allowing the meal to rehydrate, each person would taste it and grade it. If they thought it was very tasty, then they would bring the meal to my father for a tasting. If he liked it, then he added the meal to his own possible-selection list for the flight.

Most everyone had heard about the freeze-dried space food, but here were boxes of meals spread out all over the kitchen counter and dining room table.

Everyone was sharing different tastings with each other, and it was a very unique party. Toward the end of the party, the backup crew of *Apollo 14*— Gene Cernan, Ron Evans Jr., and Joe Engle—went around the house pasting up the *Apollo 14* backup patch. I was standing in my parents' bathroom in front of the sink, when Gene came running in and stuck one of the patches on the inside lid of their toilet. The backup patch, instead of the astronaut symbol, had a picture of the Road Runner from the *Looney Tunes* cartoons. He was redheaded to symbolize my father.

As the launch date of *Apollo 14* grew near, we moved down to Cape Kennedy. *Apollo 14* was a first for crew quarantines. Because Charlie Duke's kids exposed Ken Mattingly to the measles virus, Mattingly was pulled from launch two days before *Apollo 13* and substituted with Jack Swigert Jr. So for the first time, NASA decided to institute a preflight quarantine from the astronaut children for three weeks before liftoff.

We could talk to our father through a plate glass window. When they were at Cape Kennedy, wives were allowed to visit. My mother would visit, leaving us in Houston with babysitters. One day, after we arrived at the Cape, our father requested our presence, and two people from the NASA public relations team came out to pick us up. Only, we weren't where they thought we were; we were at the beach with our cousins. I think they contacted the Cocoa police and a full APB went out on us over the police radio system. When they finally found us, we were taken in our swim gear with a police escort to visit with our father.

On launch day, 31 January 1971, we were out at the NASA viewing area. Only, this time, our families were the VIPs. After the astronauts suited up, they proceeded to the astronaut transfer van. At the exit door was a crowd waiting to have a last glimpse of the crew before they boarded. Major Stuart Roosa asked NASA to mark a spot for his parents and his brother's family to stand. Dutifully, NASA had cordoned off an area with a red square carpet and velvet rope stanchions. My grandparents were there along with our uncle Dan and his family. Daddy told us, "I didn't want you there because I thought it would be too emotional."

A NASA spokesman came over and politely asked my grandparents if they would share their area with two other NASA VIPs. My simple grandparents said, "Of course, anything for NASA." They didn't recognize the two VIPs: Henry Kissinger and Charlton Heston. As my father walked out, he reached out and touched his parents.

I think part of his desire to have his parents there to see him off was that my grandmother always put my father down for leaving "God's Country" (Oklahoma) to pursue an air force career. I think my father wanted to touch them to show them how far he was getting ready to travel—to the moon.

As we came closer in the countdown, there was a delay for weather. There were thunderclouds over the Cape, and NASA didn't want a repeat of *Apollo 12*. At the moment of liftoff, *Apollo 12* was hit by lighting, causing the engines to shut off and restart, all in a millisecond, but it could have been a disaster.

My father said that when the count restarted, all he could see were the dark clouds out the window. But the rocket trajectory had them bypassing the cloud. The countdown continued: ten, nine, eight, seven—the ignition sequence started with the flames of the Saturn V first stage—five, four, three, two, one. They had liftoff! I shouted as loudly as I could, "Godspeed! Godspeed! Godspeed!" over and over again. The roar of the thunderous noise passed through my body, drowning out my shouts. The rocket cleared the tower and passed up through the clouds. I was still yelling when the rocket went out of sight.

We were taken back to the crew quarters and transported to the airport to get into our friends' private jet to return us to Houston. As we boarded, a NASA representative told my mother, "They can't get the command module to dock with the LEM." As we got on the plane, I said a prayer for my daddy. When we landed in Houston, my middle brother, Jack, was asleep, and a close family friend carried him off the airplane. A picture was taken that went worldwide that moment. When we landed, we were told that the crew had completed the docking and were now safely on their way to the moon.

It took them six attempts to get it docked. At one point, my father hit the LEM so hard that it shook. Alan Shepard said, "Be careful, Stu; don't run through it." He would later say that Alan Shepard was trying to figure out a way to do a space walk to get it docked. My father decided to take one more stab at it; he backed off, looked at the window, and said a Hail Mary. He came in again, and the latches captured for a hard dock. Daddy believed in the power of prayer.

The next morning, when we got up as usual for school, my mother was making breakfast and asked me go out and get the newspaper. It was still dark when I stepped out of the front door only to be met by a horde of photographers jumping out of the bushes. I was so taken aback that I stepped back, shocked, and then I just turned around and walked back inside without the paper. I told my mother, "I forgot to get the paper," and told her what had happened. We were not prepared for that type of paparazzi.

She sat me down, and we ate breakfast and went to school. Of course, my fellow students and friends knew my father was on the way to the moon, but it was not really that special any more. There were plenty of other astronaut kids in the school whose fathers had already flown and other kids whose fathers were going to fly next. Many of the other kids' fathers were scientists, engineers, or worked for NASA in some capacity.

The afternoon of the second day, my brother Allen was riding his bike and saluted the photographers, with his wrong hand, but it again made the AP wire. It was published everywhere. I was getting jealous that my brothers were getting all the news.

On day three of my father's mission, we were told we were going to be on the CBS *Evening News*. NASA drove us out to a spot that had the space center in the background, and we were sitting on the edge of a large water fountain that had been turned off. There was a TV camera set up, and we were all told to sit down. We sat down in order of age, which meant I was sitting next to my mother.

They put an earpiece in my ear, and suddenly I heard voices. No one had briefed me on what was going on or what I was supposed to do. I suddenly heard this voice ask my mother a question, and she started talking. I was trying to figure out where the voice was coming from. Then this baritone voice asked me a question. I looked at the camera and responded. I was nervous, I was confused, and then it was over.

The NASA folks came up and said, "Great work." I was then informed that I was the first astronaut kid ever to be interviewed by Walter Cronkite, live on the CBS *Evening News*. I got home, and everyone was saying what a good job I did. I was disappointed that we didn't have some way of recording it.

Inside our house, we had two squawk boxes, small metal boxes with speakers receiving a live feed from mission control. We could listen to the CAP-

COM speaking with the crew. We had one in our family room and one in my parents' bedroom.

I can remember a time early one evening when it was dark out. I had the back door open, standing on our patio, looking at the moon, and the CAPCOM had called the house and was talking to my mother. My mother was relaying how we were getting ready for dinner and that she had cooked lasagna, one of Daddy's favorites. The CAPCOM relayed the message, and my father responded that he was just getting up and making some orange juice. As I listened to this exchange, I stared as hard as I could at the moon, knowing that was where my daddy was. I believed if I looked hard enough, I would see his capsule. Needless to say, I didn't. It was a surreal experience to be listening to my father's voice live from the moon.

During the flight, *Apollo 14* was the first mission to carry a high-contrast, or high-con, camera. My father was told that it cost one million dollars. He was to use it to take photographs of the *Apollo 16* landing site, but the camera wouldn't operate. He had a long discussion with mission control troubleshooting the problem. Nothing worked. Daddy had an expression, "If something doesn't work, hit or kick it hard enough, and it might fix it." He said, "I thought about banging it real hard and flinging it around the cockpit. But with it costing a million dollars, I didn't want to damage it." Once they returned, the technicians discovered that a tiny metal sliver had floated up in the gear and blocked it. He said, "If I would have whacked it, it might have loosened up the sliver, and the camera could have worked. Should have gone with my gut."

Instead, my father mounted a 120 mm Hasselblad camera into the window. To take photos of the landing site, he had to fly the command module with one hand and point the nose toward the landing site, thrusting to have the command module do an arc over it. Meanwhile, he used his other hand to work the shutter of the camera at a constant rate. He was later told the pictures were exceptional. He was proud of that.

He told us a piece of photographic trivia. When the crew returned, the negative film was put into a machine that would slice a very thin layer off the entire strip of film. This test strip was then developed to make sure the developing times were correct. NASA only had one shot at making sure they didn't destroy an entire roll of negatives with the wrong chemicals or development times. I was amazed that someone had thought of doing that.

My father talked of what it was like to first be in weightlessness. He said, "It felt like you were standing on your head. Your heart is so used to pumping against gravity to get blood to your brain, that it takes some time for the brain to tell the heart, you don't need to pump so hard. After a while, it goes back to normal."

He also talked about what it was like trying to sleep in space: "I was used to resting my head on a pillow, and it was hard to try to sleep with your head just floating. I would float under the center seat and use the bracket rod to wedge my head into it to make it feel like it was being held down."

We had a full-time NASA public relations individual assigned to the family. He did a great job. All the astronaut wives shared stories of dealing with the press. My mother knew the journalists all had deadlines, so she would hold a press conference early, at midday; answer their questions; and let them get their photographs. Then they were off until the next morning. After the miraculous recovery of the *Apollo 13* crew, the press was all over the families of *Apollo 14*.

On the day of splashdown, my mother had a TV moved out into our garage for the press to watch. We had some of the local TV crews inside the house to record the family's reaction to seeing the command module floating down into the Indian Ocean. We had our family priest there, along with other close family members and friends. Seeing the parachutes, we all cheered.

There is a dot on a map that marks the splashdown site for each Apollo capsule. Daddy made the comment, "I think I will just land it on the carrier." Apparently, unsure if my father was joking, the primary recovery vessel, the USS *New Orleans*, was moved an additional five nautical miles away from the splashdown site. In the end, he missed his splashdown site by an eighth of a mile. That was the closest splashdown of any Apollo mission. Al Shepard missed his spot on the moon by seventy feet, again the closest of any lunar landing. The other thing my father was proud about was that they didn't have to inflate the balls on the top of the capsule. If the capsule landed inverted in the water, the balls would inflate and right the capsule.

On the TV, we watched as the U.S. Navy SEALS put the inflatable collar around the capsule. Al Shepard was the first out, since he was riding in the center seat. Next was Ed Mitchell. Finally, my father emerged; he rubbed his hand on the capsule as a symbol of thanks. They were then taken into a basket and hoisted up into the helicopter.

We watched Daddy walk out of the helo on the hangar deck of the USS *New Orleans* and into the Airstream trailer. *Apollo 14* was the first crew to have preflight quarantine; it was also going to be the last flight to have postflight quarantine—three weeks of being sealed off, with a few doctors and debriefers, to make sure that no virus like that in the movie *Andromeda Strain* was coming back from the moon to infect Earth-bound mankind. My father was always mad that when he was weighed on the ship, the doctors recorded him as having lost weight. He would say, "I didn't lose weight. The ship was rocking, and the scale numbers were moving back and forth."

We were at Ellington Air Force Base, located near NASA, when the crew arrived in the Airstream trailer and unloaded onto a tractor-trailer rig. We waved through the large window on the rear of the trailer and briefly spoke to them via a telephone handset. It was a bright day, and there was a full crowd to welcome them home.

So for the next three weeks, my mother would drive us out after school, and we would talk to our father behind the glass window. Sometimes we would just rather stay home to watch TV and let our mother go by herself. It was boring to talk to my father behind the glass. We all had short attention spans.

On the first night Daddy was home out of quarantine, I said, "Everyone is saying the mission went well," and I asked him whether he thought his mission was a success. "We did good," he said. Little did I know that shortly thereafter he would be on the *Tonight Show Starring Johnny Carson* and would relate my comments to the TV audience.

1. Uncle Danny, age five (*left*), and Stuart, age three (*right*), at Yellowstone National Park. Roosa family archive.

2. Unidentified individual and Stuart with his dog, Skippy. Roosa family archive.

3. Stuart (*left*) and unidentified individual (*right*). Roosa family archive.

4. My parents' wedding. Joan and Stuart (*center*); to my mother's right are her two sisters, Pattie (Barrett) White and Gloria (Barrett) Reese. On my father's left are his brother, Dan, and my mother's brother, John T. Barrett. Roosa family archive.

5. Out celebrating a traditional Japanese meal during my father's tour at Tachikawa Air Base near Tokyo; I'm in the middle, with my brother, Jack, and my parents. Roosa family archive.

6. Family portrait taken in 1962 with the addition of Allen (*far right*). Roosa family archive.

7. Stuart poses in front of an F-104 Starfighter at Edwards Air Force Base, California. Courtesy of
Stephanie Smith, USAF, Air Force Materiel Command.

ROOSA STUART A
MAJ   FR50873
10 MAY 67

8. My father in his major's uniform in 1967. Courtesy of the U.S. Air Force.

9. The fifth selection of NASA astronauts, the "Original 19," as they called themselves. My father is fourth from the right in the back row. Courtesy of NASA.

10. Part of astronaut preparation was jungle survival training. *From left*: Stuart, Ken Mattingly, and Al Worden. Later, they would all fly in Apollo as the command module pilots of *Apollo 14, 16*, and *15*, respectively. Courtesy of NASA.

11. Training for splashdown recovery. *From left*: Alan Shepard, Stuart Roosa, and Ed Mitchell. Courtney of NASA.

12. Mitchell, Shepard, and Roosa pose in front of the *Apollo 14* Saturn V as it slowly rolls from the Vehicle Assembly Building to Launch Pad 39A on 9 November 1970. Courtesy of NASA.

13. My father's parents, Dewey and Lorine, peer into the *Apollo 14* command module.
Courtesy of NASA.

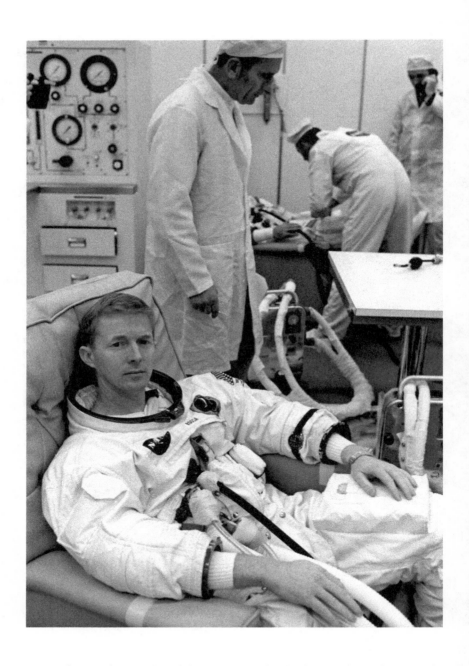

14. Just a few hours from launch, my father looks so relaxed in this photograph. Deke Slayton is in the background. Courtesy of NASA.

15. The *Apollo 14* crew in a jovial mood during a photo shoot. *From left*: Shepard, Roosa, and Mitchell. NASA photo courtesy of Spacefacts.de.

16. *Apollo 14* at liftoff. Courtesy of NASA.

17. *Apollo 14* at liftoff. Courtesy of NASA.

18. Launch day at the astronaut viewing area, the family dressed in red, white, and blue. *From left*: me; my sister, Rosemary; my mother, Joan; and my brothers, Jack and Allen. Roosa family archive.

19. Launch day photo op of the wives. *Apollo 14* was the first Apollo flight to have all three wives present. *From left*: Joan Roosa, Louise Shepard, and Louise Mitchell. Courtesy of NASA.

20. My father flying solo around the moon. The command module, *Kitty Hawk*, and the service module. Courtesy of NASA.

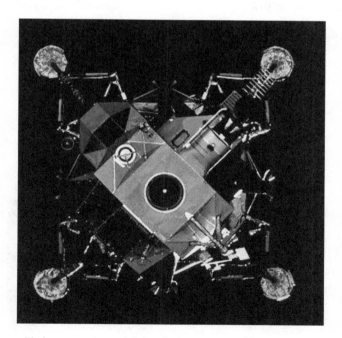

21. The lunar excursion module (LEM), *Antares*, separated from the command module to begin its descent. Courtesy of NASA.

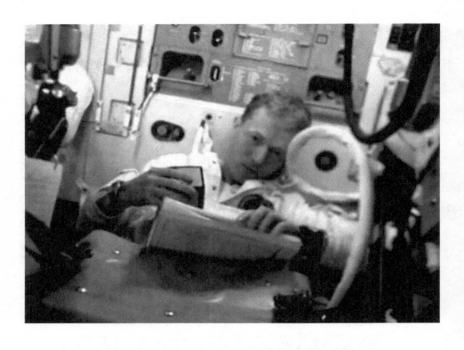

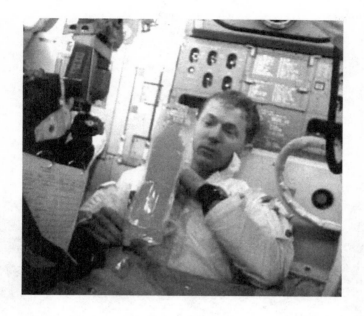

22. (*top opposite*) My father looking at the checklist during his mission. Courtesy of NASA.

23. (*bottom opposite*) Wearing his eye patch, my father getting ready to use the sexton to do star sighting for navigation. Courtesy of NASA.

24. (*above*) My father drinking orange juice. Courtesy of NASA.

25. Splashdown. Shepard (*foreground*) and Roosa (*background*). My father was extremely happy that when he landed, he did not have to use the inflatable balls to right the capsule. Courtesy of NASA.

26. (*top opposite*) Back on Earth. The crew in the quarantine trailer for transport back to Houston. *From left*: Roosa, Shepard, and Mitchell. Courtesy of NASA.

27. (*bottom opposite*) Speaking at a press conference. *From left*: Shepard, Roosa, and Mitchell. Courtesy of NASA.

28. My father receiving back from the Forest Service the container that he used to carry the moon tree seeds aboard the command module. Courtesy of NASA.

29. Stuart Roosa Day in his hometown of Claremore, Oklahoma. The elementary school was renamed for him. Roosa family archive.

16

Moon Trees

One of the legacies of my father's moon mission is the so-called moon trees. These are the tree seeds that traveled with him on *Apollo 14* as part of his personal kit.

My father wanted to honor the U.S. Forest Service because of his days as a smoke jumper. He spoke to the chief of the Forest Service at the time, Ed Cliff, and the idea of taking tree seeds to the moon was born. Five different types of tree seeds were selected: loblolly pine, sycamore, sweet gum, redwood, and Douglas fir.

After *Apollo 14*'s flight, those seeds were germinated in Houston. Apparently, over a three-day holiday weekend, the power went out in the greenhouse. The greenhouse overheated, which killed many of the saplings. A caretaker came in over the weekend and realized the issue. He loaded up the remaining saplings and drove them to Gulfport, Mississippi, to a U.S. Forest Service facility there. This saved many of the trees. In 1976, the bicentennial of the United States, many moon trees were planted across the United States and abroad. My father flew all over the place planting trees. This was not a formal NASA experiment or endeavor, just my father's quiet way of honoring the outdoors that he so loved.

17

The Country Western Tapes

Daddy loved country music. Over the course of time, he and Bill Bailey, who was the prime-time DJ for Houston's biggest country station, KIKK, became friends. For the Apollo missions, they had a small cassette recorder on which they would make homemade tapes so the astronauts could listen to music on their flights.

Bill Bailey decided to do something special for my father. He used his contacts in Nashville to have some of the country stars of the day cut a special tape for Stuart Roosa. Mr. Bailey contacted Sonny James, Jerry Lee Lewis, Buck Owens, and Johnny Cash to make personal messages and songs.

Bill took those tapes, wrapped them up, and handed them to my father to place in his PPK (personal preference kit). Each astronaut was allowed a certain amount of weight to carry personal items. The tapes were already locked inside the capsule when Johnny Cash's manager made a public announcement that Johnny would be coming out with an album, *Johnny Cash to the Moon and Back*.

NASA officials were upset to learn that someone was going to try to make money off their mission to the moon. They asked the astronauts if they knew anything about it. Daddy said that he had been given some tapes but that he didn't know what was on them. NASA technicians went into his PPK, found the Johnny Cash tape, removed it, and passed it to my father. So he launched without the Johnny Cash tape.

My father talked about the first time they were doing a burn to put the command module and lunar module into lunar orbit. As they were going into orbit, he put the Sonny James tape in the recorder, and the astronauts listened to "How Great Thou Art" and specifically the first verse. "I see the stars . . . Thy power throughout the universe displayed . . ." It was the perfect song at the perfect time.

A few months later our family was at the Houston Livestock Show and Rodeo, inside the Astrodome. We were standing in a holding room, and I was talking to someone, when I heard this unmistakable baritone voice say, "Hi, I'm Johnny Cash." I turned around and was standing next to my father when they were introduced. I shook Mr. Cash's hand.

Daddy said, "I am sorry for what all happened about your cassette."

Johnny looked at my father and said, "Have you listened to it? I worked hard on it. It's a good tape."

My father replied, "To tell you the truth, we have been quite busy and just haven't had time."

"Try and listen when you get a chance," Mr. Cash replied.

Daddy said, "Will do," and that was the end of the meet and greet.

The tape lay around my father's closet, and as far as I know, it was never played.

Several years later, again at the Houston Rodeo, my father was brought onstage at two different concerts: Sonny James and Jerry Lee Lewis. And he presented them each with flown flags from *Apollo 14*. He told the crowds at both concerts the history of the personal tapes he had taken to the moon.

18

Postflight

It's important to remember that the Apollo flights were largely a product of the Cold War, and the United States made good use of their success in landing on the moon in domestic and international spheres.

For the crew of *Apollo 14*, this included a whirlwind tour of Washington DC, Chicago, and New York City. Happily for us, President Nixon sent down *Air Force 1* to take the families of the astronauts to Washington DC for the start of a tour. We landed in DC and were whisked away to the Capitol building. There, the crew gave speeches to both the House and the Senate. We sat up on the dais to the left of the podium, where the president delivers the State of the Union address. We were told we were the first kids ever allowed to sit on those seats on the House floor for this type of gathering. My father gave a very moving speech about our American flag traveling west all the way from the Eastern Seaboard to the Pacific on Conestoga wagons, then on airplanes faster than the speed of sound, and now the astronauts were privileged to carry and plant the American flag on the moon.

Meanwhile, during the speeches of the two other members, my brother Allen started jabbing me in my leg. I knew he was doing it just to tick me off, but he was bored. He also had to know that at some point payback would come. My mother had a hard and fast rule: "Do not ever misbehave in public, or you will get the belt." Other young astronaut kids were caught embarrassing their parents. Of the kids from *Apollo 14*, we had the youngest family, ages seven through eleven. Ed Mitchell's two daughters, Karlyn and Elizabeth, were in their early to mid-teens. Alan Shepard's daughters were Laura, Julie, and Alice. Laura, the oldest, was twenty years old, with Julie and Alice close behind.

After the speeches, we moved into the crush of the hearing room for the House Science, Space, and Technology Committee, with the mad scramble of politicians who wanted to shake the crew's hands. We then went to the Senate side to shake senators' hands. The only one I remember was Ted Kennedy, since I knew who he was.

The kids were then separated and taken for a tour of the White House while the crew and their wives went off to do a press conference. I remember, as we entered the Oval Office, our tour guide was not expecting the exuberance of the Roosa kids. I ran and jumped in the president's chair, pulled open a drawer on the side of the desk, and quickly opened the silver pencil box on the top of the president's desk. It was cool because when you opened it, it played "Hail to the Chief." Our tour guide was aghast. At some point, we were corralled and did a tour of the remainder of the White House. With the tour complete, we returned to our hotel, the Washington Inn. President Nixon was going to promote Alan Shepard to admiral at a state dinner that night. At some point, Alan Shepard had asked the president if his eldest daughter, Laura, could attend. At that time, protocol was that you had to be twenty-one years old to attend a state dinner. Nixon's reply was, "Let them all come!"

It was common that after a spaceflight, if you were in the military, you would be deep selected, ahead of your peers, for your next rank. My father was already a lieutenant colonel selectee, so the U.S. Air Force just moved forward his promotion date.

We returned to the hotel with our invitations in hand. At first my mother was confused, because the kids of all the previous crews had eaten dinner in the White House kitchen. Meanwhile, while my mother was sorting that out, I launched into punching Allen, as payback for poking me in the leg at the Capitol. My mother yelled at me to stop, and I did. But then I told her what Allen had done. I offered her a choice: either we behave out in public and she lets us vent when get back to the hotel, or else we would misbehave in public, because I would not be putting up with it. The rule then became that when we returned to a hotel, the first five minutes was a free-for-all.

Meanwhile, a separate panic attack had set in. My mother had not packed for all of us to go to a state dinner. No one could find the luggage that we had placed on *Air Force 1*. I remember Daddy yelling, "How in the hell do you lose luggage coming off of *Air Force 1*?" So none of the families had anything to wear. It was turning into evening, and the NASA representatives were scrambling to find a clothing store that could quickly provide tuxes, evening dresses, and kids' dress clothes. The NASA representative figured out that the driver had not been told where we were staying, so he had returned to NASA Headquarters with our luggage to wait for further instructions. He was quickly dispatched to the hotel to deliver our luggage. Crisis averted!

We were all running around trying to get dressed. My mother yelled at me, "You can't wear red socks." I had to go find another color. Ever the crisis manager, my mother had purchased a new tux cummerbund for my father. When she took out the cummerbund from the box, they realized it was defective, with two female ends and no male. My mother quickly grabbed the hotel sewing kit and literally sewed the cummerbund around him.

Right before we departed our rooms, my mother gave the etiquette brief: which fork to use, napkins, how to serve yourself, and rule number one—if the president comes to your table, you *better* stand up. I watched as Daddy leaned in so that he would be caught up on the brief.

We rode in black limousines the short distance to the White House and entered on the north side. We were herded off to a holding area, while the crew and their wives went upstairs to have a drink with the president in the private quarters. At some point, while we gathered in the lobby crowded with the VIPs, the president, the First Lady, and the crew and their wives, all descended the stairs and paused while "Hail to the Chief" was played.

The honored delegation moved to a receiving line, and we were taken to the East Dining Room to our table. We were going to be the first people under twenty-one to attend a state dinner. My sister set the record for being the youngest. She was only seven. Our table was set for ten—the three Shepard daughters, the two Mitchell daughters, and the four Roosa kids. The host for our table was Tricia Nixon.

My place card had me sitting next to her. She was beautiful in a white dress. She told me that the press was trying to figure out if she was engaged to Bobby Cox and how she would place rings on all her fingers and say, "Guess." But she essentially told me she was engaged, trusting me with her secret.

At one point, the waiters were coming around serving food. They would stand slightly behind you and serve from the left and remove the plates from the right. The food was on a large silver tray, and a large fork and spoon were on the tray. We were expected to manipulate the large fork under the food item and then with the same hand put the spoon on top of it to hold it down and lift all of it, one-handed, to our plates. It was hard to do with little kids' hands.

At one point, I dropped a piece of cheese, and Tricia leaned over and whispered, "Don't worry—it will be gone before you know it." I looked down, and it didn't exist. Tricia laughed. According to Tricia, when one of the wine glasses at the adult table was knocked over, it created a domino effect with all the other set glassware, embarrassing everyone at the table. For my part, I made sure to watch her closely—what she did, I did. That was my etiquette protocol.

At some point, dinner stopped, and the speeches started. The president called Vice President Agnew to the stage. The vice president had recently been playing golf and hit a spectator. Golf was the theme for the night, since Alan Shepard was now famous for hitting a golf ball on the moon. The president handed a golf ball to the VP, who then walked off stage and headed to our table, and he handed it to my sister, Rosemary.

The president then proceeded to promote Alan Shepard to admiral, and each member of the crew made some short remarks. At the conclusion, the president made his way over to our table. My mother watched in horror as we didn't stand up. She said later, "Y'all were so going to be beaten." What she hadn't seen was the president holding his hands out in front of him, motioning for everyone to remain seated. The president came over and shook all our hands and returned back to the head table. My mother was fuming.

My mother would later tell the story of going to the dinner, and as they approached the head table, President Nixon reached over and held the chair for my mother. She started to sit and then paused. The president asked whether something was the matter. My mother replied, "I just don't know how to take it in—a small-town country girl from Mississippi, and the president of the United States is holding my chair." He laughed, and she sat down.

At the conclusion of dinner, as guests were standing up to depart, my mother made a beeline to our table. She sternly asked, looking at me, "Why didn't you stand for the president?"

Laura quickly spoke up, "The president was motioning for us not to stand." My mother processed it and then coolly walked off.

The U.S. Marine Corps' President's Own band played music in the foyer. Some people were dancing, and a woman who was a young reporter kept asking me to dance. She clearly wanted to say she had danced at the White House. I should have taken her up on it, but I didn't want to get into trouble by doing something that violated protocol. In hindsight, my mother would definitely have encouraged me.

So it ended, the first astronaut kids to attend a state dinner at the White House.

The next morning, we were off to Camp David for three days. The president was making up for all the time the families were in quarantine. We had a police escort to take us to a helipad. I don't know if we went to Andrews AFB or Bolling AFB; however, *Marine 1*, the president's helicopter was there to fly us to Camp David. It was the first time I had been on a helicopter.

We arrived and were escorted to our cabin. Our cabin was named Witch Hazel. We were informed that it was one of the original cabins at Camp David. It is the nearest cabin to the Aspen Lodge, the president's cabin. Witch Hazel had undergone several renovations. President Kennedy turned the cabin into a nursery with rooms for young Caroline, John Jr., a nanny, and a playroom. President Johnson then modified the cabin to make it suitable for his two teenage daughters. President Nixon had it renovated in the early 1970s. It was a small cabin.

My parents had the master bedroom. My brothers and I were in one bedroom, and a there was a makeshift cot for my sister. We had a brief tour of the compound and dinner. We learned we could order first-run movies from Hollywood, and they would be delivered the next day. Or we could order from the current stock of movies on hand in the continually changing movie library. In our cabins were Camp David jackets to fit each one of us that we were allowed to keep.

Overnight Camp David was covered in a nice layer of white snow. Growing up in Houston, Texas, we had not really been around much snow. It was a thrilling experience.

I don't know how it happened that the family went in different directions. I think my brother Jack and I wanted to go out and play in the snow. We played in the snow, and then Jack and I ended up walking together around the compound. We saw a sign that said "Bowling Alley." There was a navy guy walking by, and we asked, "Can you open the bowling alley for us?" He got on the radio, and pretty soon the doors were opened. Apparently, President Nixon had instructed the commander of Camp David to take care of everyone.

My brother and I were bowling and trying out different weights of bowling balls, when suddenly, Jack dropped his bowling ball and quickly tried to get his hands underneath the ball to catch it. The weight of the ball just smashed his finger into the bowling lane. Jack started crying from the pain. Blood was coming out of the tip of his middle finger, and I was trying to console him. Whoever was watching us suddenly got on their radio, and there was lots of chatter. I am sure they were saying, "Astronaut kid, hurt at bowling alley, needs medical assistance. ASAP!"

The next thing I knew, all kinds of navy folks surrounded us. The base commander was there. Lots of discussions were going on, because Camp David did not have an X-ray machine. They were trying to find a way to determine if the finger was broken. Someone came up with the idea to take Jack to the dentist's office on-site and use the dental X-ray machine to determine if his finger was broken.

Meanwhile, there was a search on to locate my father and get him to Jack. That morning, he had taken off in a different direction, because he was told that there were snowmobiles available. He had always wanted to drive one and took one out on the trails. What he later said he didn't know was that all the trees off the trails had metal guy wires hooked to the trees to help them to grow straight. My father decided to go off the path and ended up with the snowmobile all tangled in wires. He finally got the heavy machine untangled and back on a path when navy personnel finally located him.

He was quickly rushed to the dentist's office. I just remember that the dentist's office was packed with senior management and several corpsmen, the dentist, and me. Jack was sitting in the dentist chair when they took the X-rays of his finger.

Suddenly, Daddy walked in, and everyone stood at attention as my father walked up to a whimpering Jack. His finger had already been bandaged up. Daddy looked at me and placed a hand on Jack and asked what happened. Jack was weeping, so I told the story of how he dropped the bowling ball and tried to catch it.

My father softly told Jack, "Next time, just let the bowling ball hit the floor." He hugged Jack, who stopped crying, and he walked out the door and got on the snowmobile and drove off. This time, he stayed on the paths.

I remember the U.S. Navy base commander looking at me, saying, "Is that it?" It was like he expected my father to announce that they would all be court-martialed for his son being injured. I looked at him and said, "That's it." The look of relief of everyone in the room was clearly visible. Jack, with a bandaged finger, and I walked back to our cabin.

That night, the crew and their wives went off for a special dinner. The kids were in the president's large cabin and again had a formal dinner, still trying to master the one-handed fork and spoon. After dinner, one of the Shepards' daughters ordered up the movie *The Bird with the Crystal Plumage*.

We had popcorn served in large silver bowls with the presidential seal on the side. We were about halfway through the movie when our parents returned. My mother asked what movie was on. When we replied *The Bird with the Crystal Plumage*, it turned out to be an R-rated movie. My mother's next words were sternly, "Roosa kids to bed." We grudgingly got up and walked back to our cabin.

I remember the master bedroom of our cabin; they had an ice chest full of Schlitz beer. My father had a legendary reputation for being a beer drinker. I know the other crewmembers drank scotch and bourbon, but no one knew how much beer my father could drink. Daddy was told that they would constantly refresh the beer in the ice chest, and they had an additional eight cases in the closet and a helicopter on standby if needed. When my father found out about these plans and the president's directions to keep everyone happy, he replied, "I like to drink beer, but everyone relax. I am not going through eight cases in three days." Daddy did admit that he thought about bringing a Schlitz beer into space and opening it during the flight; however, he was concerned about how it might spew inside the capsule and damage the consoles. So he decided against it.

The next day, we packed and got onto *Marine 1* again and flew to Andrews AFB, where we boarded a NASA jet to take us to Chicago.

We arrived in Chicago to a ticker tape parade. We rode in convertibles, sitting on the back of the trunk. I remember the fire tugs shooting water out of their hoses on the river. The parade ended at a convention center. There was a crush of people, and we went into an office and met Mayor Daley. We were given Cubs jackets. We then had lunch with all kinds of celebrities. I remember Jimmy Durante making fun of his big nose. We were again at a kids' table, this time on an elevated stage. The man sitting next to me was one of the U.S. Navy SEALS who had attached the float collar to the *Apollo 14* capsule on splashdown. My mother was nervous again about us kids being in front of a huge crowd and making some type of mistake.

At one point, we were escorted to the Palmer House hotel. We were given the penthouse. Alan Shepard was originally slated for it, but he generously said that the Roosas should get it because we had more kids. The penthouse was bigger than our real house in Houston. We ran from room to room exploring the place. We again had to quickly change and were off to the Ice Capades, where our families were the guests of honor.

The next day, I don't recall what event we were attending, but Allen started his antics again. I gave him, "First warning." He continued, knowing that I would give him a total of three warnings. More jabs to my leg, "Second warning." More time passed, more jabs, "Third warning." We got back to the penthouse, and I again wailed on him. He took off running and hid.

The kids were scheduled to link up with the crew and wives for another event. The security detail was rounding up everyone, and Allen couldn't be found. Mild panic set in. I was like, "Let's just leave him." Naturally, the security guys said, "We can't show up without him." We were running late, so I said, "I will help you find him."

I knew Allen's tactics. I walked around the penthouse, and inside, there was a huge saloon-style barroom. I started opening the cupboard doors under the bar, and lo and behold, there he was curled up hiding. The security guards grabbed him and carried him out, and we were on our way to whatever event we were scheduled to attend. The next day, the kids boarded the NASA jet and returned to Houston. The crew and their wives went on to New York City.

For us kids it meant back to school, right where we had left off. Our grandparents were there to babysit us, and our parents returned a few days later.

We heard that my mother and father went to the Muhammad Ali versus Joe Frasier fight. My mother told a story of meeting Mayor John Lindsey and Governor Nelson Rockefeller. Someone told her that the difference in wealth between Mayor Lindsey and Governor Rockefeller was that if Mayor Lindsay walked into a room and dropped his coat, he would turn to make sure someone picked it up, while Governor Rockefeller would just keep walking, *knowing* that someone would pick it up.

At some point, my parents left again as their NASA tour continued. They went to Las Vegas for a few days and returned. I was sitting at our breakfast table, and my mother was talking about being at the Elvis concert. I asked her if she had grabbed one of the scarves that I had seen Elvis wipe himself with on TV and throw into the audience. My mother looked at me with complete disgust and said, "He was tacky then, and he's tacky now. I would no more want a sweaty scarf than anything."

The crew and their wives then left on a NASA European tour. This was the height of the Cold War; America had won the race to the moon. Whom better to parade to world leaders than the astronauts themselves, to remind everyone that the United States had beaten the Soviet Union? My mother related a story about going to attend a dinner at Versailles. The day's schedule was full of press conferences, speeches, and other VIP events. Due to the tight schedule, the astronauts were given a police escort from their hotel to Versailles.

At some point, my father walked back to the kitchen in Versailles, where the police officers were hanging out, and thanked them for the escort. They were appreciative and were drinking wine. Apparently, one of them said, "We don't have anything better to do; we will give you an escort back." The way my mother would tell the story, after dinner, the police were escorting them, when they went into a roundabout and a car didn't yield fast enough. The gendarme on a motorcycle pulled up alongside the car that didn't yield and kicked the driver's door so hard that it put a big dent in the door. My mother said, "We got back to the hotel so fast we were way ahead of schedule. So we all went to the bar and had a drink."

Another time, my father attended some parties in Los Angeles. At a reception, as rumor has it, Barbara Eden of *I Dream of Jeannie* fame hit on him. Word about it got back to my mother. Apparently, there was a second such occurrence, and my mother was present to witness it. One morning, Daddy and I were sitting at the breakfast table, and my mother was at the stove cooking eggs. The topic came up, and my father was just looking down at his plate. I realized they were talking about Barbara Eden, and I looked at Daddy and said, "But Daddy, she is hot." My father mumbled something, but I remember my mother giving me a death-knell look. In hindsight, it was possibly something I should not have said.

My parents would be gone for days or weeks, depending on the events being held. My father's parents sometimes babysat us. My grandmother didn't like us boys but loved her granddaughter. Grandma would cook dinner for Grandpa and us, and it was always too early—about 5:30 in the evening. We tried to tell her that we ate dinner at 7:00 p.m., but she disagreed, saying that we should be in bed at 7:00 p.m. We insisted that we went to bed at 9:00 p.m.

If I did something that she thought was wrong, she would tell me to go get a switch from the backyard. I thought I was smart and would get something too flimsy to hurt or something that would break quickly or be too heavy for her to use. One time, my parents were on a three-week safari to Mozambique, in Africa. It was early in their trip, and Allen and I got into a big fistfight. My grandparents threatened to leave. Allen and I decided it was better to have a truce than to have our parents return and find out we were alone because our grandparents couldn't take it anymore and had abandoned us.

I am sure our parents would be just happy to come home and find out we were still alive. There was one time that my grandparents were babysitting us, and my brother Jack and I were playing tackle football in our side yard. With a sudden thud and bash, my mouth collided with Jack's head, and I cracked my two front teeth in half on his skull. For years, I dealt with the dentist trying to make my front teeth look normal and saying I was too young to get caps. I hated my front teeth. I couldn't eat anything that might put pressure on the teeth—candy apples, corn on the cob, or hard candy.

There was another time when my brother Jack was taken up to our local swimming pool. My grandmother allowed him to play in the baby pool. He decided to do the butterfly stroke, and using his legs to launch himself off the bottom, he sent himself straight into the side of the baby pool, cracking his teeth.

One time while my grandparents were babysitting, Allen got into a bicycle wreck and cracked his front teeth while my parents were away. My parents never seemed upset to find that we had destroyed our mouths while they were away. I suspect they were just happy to have someone babysit so that they could go on their travels.

Another time, Mrs. Irons, a very elderly lady, was babysitting us, probably because our grandparents refused, and Allen and I got into a fistfight in the master bedroom. Mrs. Irons started screaming, "I'm having a heart attack," her hands grabbing her chest. Since I had gotten in my whacks on Allen, I let up, and Mrs. Irons recovered.

At one point, one of our babysitters did quit. But I was sixteen then, so I was able to go to the store and buy groceries and cook food for us. I don't recall what my parents said when they returned and we told them. I think they mostly felt that everybody was alive and okay, so there was nothing to worry about then.

One of the best blows Allen ever got in was when we were playing in the backyard. Allen decided that it was time to get into a fight, and while I was turned away, he grabbed the hose and swung it with the water sprinkler attached. The sprinkler hit me hard in the face. My face was bleeding, and Allen took off running into the house.

I can remember him running down the hallway and yelling, "Mama, Mama, Mama!" as I was closing in on him. We rounded the corner into my parents' bedroom, and my mother was yelling, "What?" She was busy and half-dressed, putting on her makeup. Allen had run out of room, and I pounced. I knew I would have about thirty to sixty seconds before my mother could break it up. The one major rule Daddy had about fighting among ourselves was no punching in the face. I broke the rule on this occasion. My mother broke us up, and we both had blood running down our faces. She had to bandage us up. She never said so, but I think she sided with me on this occasion.

My middle brother, Jack, was known as the peacemaker. He generally stayed out of the forays, but I do remember one time when Allen and I were fighting in their bedroom walk-in closet. Jack picked up an M-16 water gun and smashed it on my head. I had beaten Allen down by that point and took on Jack. From then on, Jack stayed out of the way when Allen started his antics; Jack knew what was coming. My sister was no angel either; she would fight with Allen and then go screaming into the master bedroom, rubbing a wound to make it redder so that he would get in more trouble.

As brothers, we didn't always fight. One time in the backyard, we had secured a wine bottle and a cork and were using it as an impromptu rocket. To do that, we added baking soda and vinegar until the chemical reaction caused the cork to pop up high in the air. At one point, my father walked out into the backyard with fellow astronaut Charlie Duke. When he learned that we had been coming up with different mission names each time we launched a cork, Daddy said, "Why don't you name one Orion? Charlie has been assigned to the crew of *Apollo 16*, and that is the name of the LEM." We made an extra dose of vinegar and soda and launched the cork way up into the air. That's how we learned that Charlie was going to fly.

There were times on the weekends when Daddy would take us to the NASA gym. We would run all round and work up the courage to climb the thirty-foot rope. Sometimes, we would end up in the mission control building. The three of us boys would go onto the floor of mission control, sit in the various seats, and push all the buttons. Some seats were connected to pneumatic tubes that propelled cylindrical containers across the room. These we especially enjoyed. We would spend a lot of time launching them and running about trying to find the receiving desk where they would land, and then we'd send them back.

19

Tales from the Road

After the launch, my father and mother would often take off and leave us with babysitters. Sometimes this was because they had certain obligations for NASA, but other times it was simply because of the opportunities that came as a result of my father's achievements. We loved to hear the stories when they returned. People believed that if you were an Apollo astronaut, you could do anything. This meant they could find themselves in some very strange situations.

On one occasion, my parents were in Spain on a large ranch where bulls were raised for bullfighting. The host announced, "To Stu, I offer you my prize bull for you to fight and kill." My father told him he knew how to fly jets but didn't know the first thing about bullfighting. The host seemed dejected when my father declined the offer.

After the *Apollo 14* flight, he got involved in an organization called People to People Sports. Their goal was to provide sporting equipment to underprivileged kids around the world. Having an Apollo astronaut visiting the country would generally get a meeting with the leader of the country.

My father told us this story of going to the Central African Republic and meeting with the dictator Jean-Bedel Bokassa. Bokassa had been trained by the French military and rose up the ranks before he turned his attentions to taking over the country.

Traveling with my father were two good friends: Leonard Milton and Burt Klineburger. On this trip, Milton presented Bokassa with a high-powered hunting rifle. They were in the countryside on one of Bokassa's massive estates. They put up a paper target downrange, and Bokassa test-fired the rifle. He didn't have the rifle properly in his shoulder, and the recoil caused the scope to hit Bokassa's face, cutting him over his eye. Meanwhile, my father, Burt, and Leonard walked down to the target. There were no bullet holes in the target. Leonard pulled a pen out of his pocket and pierced the paper, dead center, in the bull's-eye. "What a great shot," they all exclaimed as they presented Bokassa with the target.

They continued to drive around the estate, when they came upon a pond that had mostly evaporated, leaving only a small amount of water and lots of mud. The lungfish can live in the mud during the dry season, with little to no water. Bokassa ordered his men to go into the mud and get some lungfish. Daddy talked about all these men dressed in suits getting covered with mud while using machetes to capture the fish.

Bokassa had ordered a state dinner that night in his palace in Bangui. They were running late, and instead of the dinner starting at 9:00 p.m., by the time they got dressed and changed, the party started at 1:00 a.m. My father noted that all the guests just stood around waiting. No one would dare leave and offend Bokassa.

Daddy talked of how the hallways, stairways, and ballroom were lined with military guards. He would say, "I just hope that if one of these guys decides to take a shot at the president, I am not around."

He spoke about Bokassa getting up to deliver a toast. He took his champagne coupe glass and tilted it forward so that the champagne was right at the lip. My father said, "No one listened to his speech; everyone was just looking to see if his hand would shake and spill a drop. He didn't. He was showing everyone in the room—he was the most cool, calm, collected guy in the room." At dinner parties, at our house, we would attempt to do the same with our water glasses and give a toast.

After the state dinner, my father ended up in a Johnny Walker scotch drinking contest with Bokassa. He didn't like scotch but said he wasn't going to let Bokassa beat him. The next morning, he had a monumental hangover, but since he had another event with Bokassa, he promptly rose and dressed anyway. When Bokassa didn't show up, my father knew he had won. Astronauts are competitive guys. A couple of years later, Daddy received a huge glossy invitation to attend Bokassa's coronation as emperor of the Central African Republic. He didn't attend.

In yet another example of the unrealistic expectations of astronaut prowess, my father was invited to be the master of ceremonies at a rodeo in The Dalles, Oregon. The plan was for several of the barrel racing ladies to come out full throttle holding the American flag, the state flag, and several other flags, as they careened around the arena, and then to suddenly pull on the reins of the horses so that they would come to a sliding stop in the middle of the arena. My father was then to ride out to the center, and the crowd would rise for the national anthem. Daddy commented, "Those girls were so smooth in the saddle, riding with one hand on the reins, the other hand holding the flag pole, riding at full speed."

He was given a large calf-roping horse. Although the stirrups were too long, my father didn't want to make them alter the saddle, so he just put his legs down alongside the horse. His plan was to go out and stop. "I grew up riding a horse," he said, "this will be easy." They positioned his horse behind the gate and made the announcement that he would be master of ceremonies. Then they opened the gate, and the calf-roping horse did what he was trained to do: go full blast out of the gate, as quickly as possible, to chase down the calf. My father's legs were flinging about; the stirrups were flapping wildly in the air. He reached the center of the arena and pulled the reins so that the horse came to a sliding stop, almost causing him to fly forward over the horse's head. "It was so embarrassing. I was just thankful I didn't fall off. Never again will I make the mistake of not adjusting the saddle." We would laugh as he would repeatedly tell us that story, and we visualized him flailing about in the arena.

On another occasion, my parents left on a People to People Sports Committee to India, Bhutan, and Nepal. They had just been to India, and their car convoy was caught up in a massive crowd parading in the streets for an Indian festival. As the cars were inching their way forward, some people in the crowd starting pounding on the car windows and doors. Others were rocking the car. My mother said, "It was terrifying; we didn't know what to do. We finally got through the crowd and back to the hotel."

My father was invited to give a speech to a large audience in Bhutan that was arranged by the U.S. Embassy. He got up to give his normal speech about his experience of going to the moon, when he received a question from the audience: "How many people did you see?" He matter-of-factly stated, "There is no one on the moon." The translator relayed the answer. My father said, "A weird murmur went through the crowd." More hands went up. "What about on the backside?" He replied, "No one there either; I did thirty-three orbits." After the translator spoke, a larger, more aggravated murmur went through the crowd. My father could sense by the way the crowd was reacting that something was wrong. So he turned to the U.S. Embassy local translator and asked, "What is going on?"

The translator replied, "They believe when they die, they go to the moon. It is their heaven. They are all wondering where their ancestors have gone."

My father was livid. "Get me off this stage NOW!" He would tell us, "Here I am destroying someone's religion and hadn't been briefed on this, and the U.S. Embassy translator is just letting me go on." Each time he told that story, he would get livid all over again.

One time, my father was in Chad, and the local escort was talking to some of the local tribal leaders about who my father was and what he had done. The escort was pointing to the moon and talking, when the chief looked at him and interrupted, "Look at it," squeezing his fingers to about half an inch apart, with the moon between his fingers, "no man can walk on something so small." He didn't understand the concept that the moon was huge but just so far away.

He went on a scimitar oryx hunt arranged by the president of Chad, François Tombalbaye. They had hunting guides and a military escort for protection in the northern part of Chad. After a few days of camping, my father saw a large oryx and shot it. Suddenly, one of the military guards took off running to cut the throat of the animal in keeping with Muslim tradition before cooking the meat. The guide starting yelling at my father, "Shoot him! Shoot him! He will slit the throat and wreck the head mount!" Daddy was not going to shoot a guy over an animal head mount.

Shortly after my father's mission, my parents were boarding a flight on Koninklijke Luchtvaart Maatschappij. KLM was making a big deal of it, since the name translates to "Royal Dutch Airlines" and the Roosa family name is Dutch. The captain invited my father to come up to the cockpit and perform the takeoff. Disappointing him, my father declined. He would say to us, "I don't know anything about flying an airliner, and I am not sure the passengers would want to have someone with no experience flying the airplane."

In 1971 my parents, along with Jim and Marilyn Lovell, traveled on safari to Mozambique for three weeks. Their guide was Adelino Pires, one of the most famous hunting guides in Africa. My mother would tell the story about a night they were in a blind near a watering hole. Behind them rose a small bluff, which shielded them from the wind. Everyone was sitting there very quietly when, suddenly, a female elephant trumpeted right over their heads. The elephant was leading her herd, and she caught the scent of the hunters just as she was coming to the watering hole. Startled, she trumpeted. Everyone in the blind rose up with their rifles, looking into the dark, trying to figure out if the elephant would charge. A charging herd would have stomped them all to death. Instead, the mother elephant turned and walked away, leading the herd away.

With a toss of a coin, it was decided that if a lion was spotted, my father would get the shot, and if a leopard was spotted, Jim Lovell would get the shot. A large leopard track was found by a killed impala. A blind was built, and one night Jim shot a large male leopard. For my father's lion, it was a different story. They were driving at night through the tall grass, when two large brother lions where spotted. He got a shot off. The lions disappeared into the tall grass. Out of the grass, a lion charged toward the jeep, and my father got off another shot. At this point, they didn't know if he had shot one lion twice or two lions once.

Adelino and my father got out and went walking into the tall grass. They found a large male lion dead. It had two bullet holes. They were happy they had not shot two. Later one night, to celebrate my father's birthday, they took an elephant dropping the size of a large cantaloupe, squirted it with ketchup, put a candle on it, and gave it to him. He laughed.

On another trip, my parents were in Tehran, Iran, heading to a state dinner hosted by the shah. My father was invited to ride with the shah himself, while my mother was to ride with the head of SAVAK, Nematollah Nassir. SAVAK was the intelligence service, secret police, and domestic security during the Pahlavi dynasty. The two cars started to race each other to the banquet site. Nassir's car was jumping the sidewalks, driving against traffic, and going at great speed. That terrified my mother, who yelled at him to slow down. "It's fine," said Nassir, "I am the head of SAVAK. I can't be arrested." Finally, my mother yelled, "I am not going to die here in this car. But if you don't slow down, I will see if I die when I jump out of it." With that, she opened the door of the vehicle, signaling that she meant business. Disgruntled, Nassir slowed down. My mother would tell us, "He didn't like being told what to do, and much less by a woman."

20

Apollo Launches

My mother was one of the few wives who went to every Apollo launch. She also had cracked the code on getting to the Cape. All the astronauts would go out to Ellington AFB and use the NASA T-38 jets to fly back and forth to the Cape. These high-performance jets would just need a couple of hours to fly from Texas to Florida.

Now, for a wife to get there, she would either fly commercial or drive. Both took too long. The right answer was to fly by private jet. In the days of Apollo, the astronauts were the biggest rock stars of the day. There were many rumors about the hordes of ladies who were astronaut groupies at the Cape, and an Apollo launch was *the* social place to be. So if you were rich and had a jet, you just needed someone with all the right access, to all the right parties, who knew all the right astronauts. Hence, the rich turned to astronauts' wives.

For one launch, my mother was flown to the Cape on the jet of the folks who owned Chiquita bananas. For another launch, she was invited to fly with the man who had invented and patented the orange cones used on interstate highways. His plane had beds for long-haul flights and even his and her bathrooms. Yet another time, my mother rode on Howard Hughes's jet with Robert Mahue, who ran Howard Hughes's operations. He was hired and fired by Howard Hughes over the telephone, never once meeting him face-to-face. My mother would get to travel to the launches in style, and in exchange, the host couple would be invited to the right receptions and meet the astronauts.

The children, too, managed to ride on some private jets. We would travel to the Cape with my mother. But we were also invited to hunt on private ranches and country estates and to visit various other places with folks who wanted a connection to NASA and the astronaut corps.

21

Astronaut Downtime

Thanksgiving was a very special time in our house. In the build up to it, if a member of the family had shot or caught anything impressive, then it would be frozen for our Thanksgiving dinner. Daddy was proud of the fact that the Roosa family had contributed to the dinner. One year, my brother Jack had caught a very nice redfish. Somehow in the preparation, it didn't get scaled. I can remember my father's frustration while trying to serve it, when he realized why he couldn't get through the skin. At another dinner, I had shot my first duck, a wood duck, and I was so proud to contribute it to the meal. It was up on the counter waiting to be prepped when our Siamese cat, Martha, jumped onto the counter, bit into it, and took off running with my duck carcass. I just happened to be in the kitchen to witness it, and I took off chasing the cat. I about choked the cat to get my duck back. Our Thanksgiving meals would include a goose, ducks, squirrels, doves, quail, bass, redfish, catfish, trout—the more exotic the better.

Easter was always a festive time at our house. Everyone would dress up and go to church. We would get home and change, and then began the ritual of the Easter egg hunt. Daddy would go out into the backyard and hide the eggs. My mother would spend the day before making several dozen hard-boiled eggs, which she dyed all different colors.

The rules were simple. The most obvious eggs, those lying in the open, were for my sister. She was the youngest, and my father wanted her to get her share of finding eggs. Then there was one egg that was always the hardest to find. Daddy took great pride in his egg-hiding ability. Some were hidden high in the bushes; some, hidden in small parts of the swing set. He would give us clues—"getting hotter" or "getting colder"—as we all hunted for the grand prize egg. One year, we spent agonizing minutes getting "hotter" and "colder" on our small back porch. We all knew we were close, but we just couldn't see it. Allen found the prize egg. Our father had put it in the dog bowl and poured wet dog food over it. Allen reached in with his hand and fetched it out. He was proud of himself that day. Other times, while playing in the backyard, we would find eggs that had been hidden for years.

With only twenty-eight astronauts in the Astronaut Office, some were good friends; some, not so much. But most admired each other's skills. At times, my parents would throw parties, and we had all these American heroes standing around in our family room. I realize the uniqueness of this now, but to us astronaut kids, they were just regular people, my parents' friends.

It was Bill Anders of *Apollo 8*, for example, who taught me to water-ski. When I was about nine years old, Daddy and I went out with Bill Anders on his ski boat. I was going to learn to water-ski. We were out on Taylor Lake, located near our house, and I was put in the water, with my two skis. I remember Bill handing me the ski line and saying, "Don't let go of the handle." I was nervous as the boat started to pull away to take out the slack. Then I heard, "Ready?" and I nodded.

The boat roared to life. I started to get up, and then I got jerked forward and was pulled right out of my skis. I was saying to myself, "Don't let go of the rope." I had water shooting up my nose. My body flailed on the surface. At some point, I couldn't take it anymore and, disappointed with myself, let go of the rope. I don't know how many seconds that moment lasted, but it seemed like a long time.

The boat circled around, and my father yelled, "Why didn't you let go of the rope when you came out of your skis?"

I replied, "I was told not to get let go of the rope."

They all laughed, "Okay, next time, let go of the rope."

They took the boat and went back for my skis as I floated in the water. They tightened down the foot holder and slid them back over the water for me to put on. I got them on—this time, with more force in my legs to push against the water. Bill hit the throttle, and I pulled myself out of the water as I stiffened my legs and rose up to stand on my skis. I was water-skiing! I was so happy. The boat turned to circle back, and I crossed the wake and wiped out again. But I had learned how to get up, and I loved it.

I also went out on my first deer hunt with Jim and Jay Lovell. Again, when I was about nine years old, Daddy got an offer to go deer hunting, and he took me along. It was a cold, dark Texas morning. We got out, and we joined up with Jim Lovell and his son Jay. Jim Lovell was famous for going to space in both the Gemini and Apollo programs. After flying on *Apollo 8*, he would later go on to command *Apollo 13*. Lovell and my father both loved to hunt. They did several safaris, along with their wives, to Africa. On this particular morning, we got dropped off at the deer blind. My father climbed up, inspected it, and came back down. He said that the blind would only hold three people, and since Jay was older, he would get first shot. I was told to be still and not make any noise. For several hours, I sat at the base of the blind, shivering and looking at some bushes without leaves. They never saw any deer, so Jay never got a shot off.

We took several vacations to the Las Brisas hotel in Acapulco, Mexico. It is built on forty acres, on the hillside of Acapulco Bay. It has over two hundred cottages, each with its own swimming pool. Every day, they would clean the pool and leave azalea flowers floating in it. Each cottage had its own white-and-pink jeep. On one trip we took there, we went down a few days before my father was due to arrive. My mother was driving down a steep, windy road and could not get the car to go into gear. We were going fast, and we kids all kept yelling, "Shift, Mama! Shift!"

She kept yelling back, "I know how to drive, and I can't get into gear!" She finally got the jeep to stop and got into first gear, and we immediately went and got a new jeep.

There were several astronaut families staying there. The general manager threw a big party down at the seaside restaurant. The evening culminated with a fireworks display, which included a massive series of Apollo patches. It was impressive.

For the return trip home, my parents, along with the Hartsfields—Hank, Fran, and their daughters, Judy and Kelly—decided to take an overnight train from Acapulco to Laredo, Texas. The Hartsfields were very close family friends. The trip was a disaster. Before getting on the train, we ate at the train station. I ordered my favorite Mexican dish, cheese enchiladas. As it turned out, the sour cream was spoiled, and I got a case of Montezuma's revenge. My mother gave me lots of Pepto, trying to make the trip more comfortable.

Once on board, we explored the length of the train. We were in the economy cars. We walked up to the front and saw the locomotive and then back through the first-class cabins, the economy cars, through the sleeping cars, and toward the back railcars filled with people sitting on the floor with their goats and other livestock. It stunk. As evening fell, we were shown to our sleeping car. There were several railcars with small bunk beds and a curtain. My parents and sister were to sleep in one, and the three boys were to sleep on a lower bed. My parents were in a separate car from us. For us, it was too crowded. Each brother tried to find a square inch in which to sleep. Judy and Kelly were on the opposite side of us; they were fighting about space and kicking each other. There was a little hammock that ran along the wall to put personal effects into for storage. My brother Allen crawled into it, giving us more space. Everyone was bedraggled as we got off the train in Laredo. We never did another train trip again.

22

Growing Up after *Apollo 14*

It was not easy to be the son of an Apollo astronaut. For one thing, as I've said, I was a small kid and didn't really grow until after I left high school. Also, others often had unrealistic expectations about what I could achieve and then questioned whether any successes I had were due to my father or my lineage.

As a teenager, I certainly got up to my fair share of mischief. No doubt some of that was rebellion, something that has led more than one astronaut kid permanently astray. But in my case, it also reflected the risk-taking attitude that prevailed in the Roosa household. Not surprisingly, my father had a high tolerance for risk, but my mother also let us go out and live our youthful adventures, with all the consequences. I reflect on this now as a father in my own right and wonder whether such an approach is even possible in this day and age? For example, my brothers and I would often spend long hours hunting and exploring in the wilderness on our own behind our house, returning only at end of day. Or we would take our bikes and be gone the entire day. It's doubtful this would sit comfortably with any responsible parent these days.

Most of our teenage antics were the usual—sneaking out at night, taking our parents' car. For instance, one night, we had a guy sleep over at someone's house. The plan was to sneak out with the parents' car at 3:00 a.m. But when the time came, I was too tired. I was the only one who didn't go. I remember waking up at 4:00 a.m. and hearing the excitement of the thrill ride. I felt bad that I didn't go. Another time, on the night of 14 April, my friend Danny was going to sneak out the family car. I was to rendezvous at 3:00 a.m. at the intersection of Repsdorph Road and Pebblebrook Drive. I quietly snuck out of our house and walked the five blocks to hide low in the ditch on Repsdorph Road. I waited and waited, and finally at 4:00 a.m., I walked back to my house and snuck back inside. The next day, the gang asked Danny what happened.

Danny's response was that his father was up all night doing the family's taxes before the 15 April deadline, so Danny was not able to get the car.

The next time, it was my turn to take the family car. I got up in the night, went into our kitchen, took my father's Blazer keys, and quietly went outside to the car. My heart was racing. I put the keys in the ignition and partially turned them to start up the generator. I put the car in neutral and let it back down the driveway into the street. Then I turned on the car and started driving. I picked up the gang, and we drove around for about an hour before heading home. I dropped everyone off and started back to our house. Our house was on a curve. My plan was to accelerate to about 40 mph, put the car in neutral, turn off the car, and quietly coast up the driveway. It was a plan.

Only, it didn't work out that way. I accelerated to 40 mph, put the car in neutral, and turned off the engine, and that is when my plan failed. Suddenly, I heard a click as the steering wheel locked up in a slight right turn. I pushed down on the brakes; with the engine off, there were no power brakes. I used the steering wheel as a lever to stand and push down on the brakes as hard as I could. It was stark terror!

With my momentum and speed, and with my right bearing, the car went right across the neighbor's yard on the inside of the curve. There were big tire tracks right across his yard. The car finally came to rest on the far side of the street. I figured I was busted, so I turned on the car, drove it up into the driveway, and parked it. I walked into the house expecting to be met by my parents.

Suddenly, I realized the house was very quiet. I tiptoed into the kitchen and put the keys away and went back to my bedroom. It took a while for the adrenaline to wear off. When I got up the next day, I walked out to go to the bus stop near our driveway. Daddy was standing outside getting ready to leave for work and talking to the neighbor across the street about "the idiot" who drove across the yard. My school bus arrived; I got on and let them keep talking. That was the only time I ever snuck out the car.

Shortly, thereafter, my friends and I all turned sixteen and got our driver's licenses. On my sixteenth birthday, I went to the Texas Department of Motor Vehicles to get my driver's license. I had been driving tractors since I was twelve. I was driving our family Chevy Blazer since age fourteen. At fifteen, at our high school in Clear Lake, we were offered driver's education in the morning for a couple of weeks to obtain our learner's permits. We could then drive on the road with a driver over eighteen. My parents let me drive quite often.

At the DMV, we arrived with our green Chevy nine-passenger station wagon. The inspector got in and started giving commands: "Go straight," "Turn right," "Go to the light and make a left." At some point, we arrived at the parallel parking test spot. It had two orange cones marking the space. It was the final evaluated event of the test. I knew that if the car hit a cone, that would be an automatic failure. With confidence, I pulled up and backed the station wagon straight back into the spot. The evaluator then struck terror into me. He said, "Wow, I didn't think this car would make it." Now I was even more concerned about trying to get out, so I did a long series of short backing up and cutting in and out until I got the station wagon back on the street. I passed and got my license.

My parents purchased me a used burnt-orange Chevy Vega. It was a small four-passenger car. My most common companion was my brother Jack, who was a freshman and rode with me to our high school.

My friend Bruce had a large muscle car with a big V-8 engine. I had my little four-cylinder engine. One night, when we came off NASA Road 1 onto Lakeshore Drive, we were racing. As we turned onto Cedar Lane, a policeman turned on his lights. The next intersection was Bayou View Drive. I broke left, Bruce broke right, and the cop broke left. The red and blue lights filled up my car. I pulled over. The policeman asked for my driver's license, and I handed him my wallet. He said, "Please pull out your license." I nervously pulled it out and handed it to him.

He wrote me a ticket for "expedition of accretion." The next day, after my father got home from work, I told him what had happened. I told him about how the policeman had refused to take my wallet. Daddy said, "That is so someone can't do this," and he pulled out his wallet with some cash sticking out and pretended to hand it to the police officer, clearly as a bribe. I was like, "Okay, never thought of that."

At first, my father thought it was a ticket for running a stop sign, but when I said it was for "expedition of accretion," he said, "You are not getting out of that one." This was my first ticket. As the court date approached, I got more nervous. Traffic court was in the evenings so that residents could go after work; my time was for 7:30 p.m. My father loaded the whole family into the station wagon, and we went to our little civil community center for our small neighborhood. We all went in and sat down. I looked over to the judge's bench, and my spirits bolstered. Behind the judge was a plaque that said, "Astronauts of El Lago." I hoped the judge would recognize the name.

At some point, I was called to the bench. In front of my entire family, the judge asked how I pleaded. I responded, "No contest." He said, "That will be a $10 fine and $2.50 court fee," and banged down the gavel. I walked over to the window and paid my fine. We loaded the whole family back into the station wagon for the short drive home. I later learned that I could have just mailed in the fine. For the life of me, I don't know why my father made it a family spectacle. Let's bring the family down and watch the oldest son face the judicial system—maybe a lesson for my siblings and me.

In the early 1970s my parents bought jointly with a good family friend, Bob Jamison, one hundred acres of land about three miles from Dayton, Texas. Bob Jamison's family was well-to-do. The Dayton State Bank had been in his family for years. In addition, they owned land with oil wells. He was a well-liked man and president of the Duck Unlimited Chapter in Dayton. Dayton, a town of less than four thousand people, could have been taken out of a Hollywood script. The corner drugstore, with its soda fountain, was out of a 1950s movie. The hardware store and feedstore were fun places to visit. Bob would take us into the bank and show us the antirobbery devices—silent alarms, the money drawer with the last bill attached to an alarm switch, and timers for the bank vault. He would challenge us to lift a one-thousand-dollar bag of pennies—it was not easy.

My mother named the new land "The Place." A paper company owned the surrounding land, so we literally had thousands of acres to explore, hunt, and fish. It was near Dead Man's Curve and about a forty-minute drive from our house in El Lago.

Almost every weekend during the school year, we would depart on a Friday afternoon and return on a Sunday evening. The Place was really a *Boy's Own* adventure land where we were often left to our own devices. I have such precious memories of these times. Upon reflection, I think our adventures mirrored the kind of experiences that my father must have had growing up in rural Oklahoma and that my mother must have had in rural Mississippi.

Bob managed to get a bulldozer, and my father cleared the dense yaupon trees on the top of the hill for a house to be built. As Daddy cleared the land, he made a huge pile of brush and debris. My mother and Bob sketched out a design for the house on the back of an envelope. Bob passed the note to the builder as the only blueprints for the home. It was to have master bedrooms in opposite corners, boys' and girls' dorm rooms at each end of the living room, two bathrooms, and a guest room. The kitchen was part of the living room. It also had a screened-in front porch. The hill dropped down to a creek, and then the bottomland would go for about four miles, ending at the Trinity River.

In most years, the Trinity would flood the bottomland, and the only roads that would be passable were the built-up roads for the oil companies to come and inspect their oil wells and storage tanks. There were lakes, creeks, and forest for as far as the eye could see. It was a kid's outdoor paradise. I would spend hours by myself exploring, hiking, hunting, fishing, and swimming.

In 1970 Uncle Bob, with my mother's permission, bought us a Honda QA-50—the smallest minibike on the market. My mother had two rules: everyone on the minibike had to wear a helmet, and you could not take the minibike on a public road. My mother had heard too many stories of motorcyclists being killed by cars. We began to explore with hours of running up and down the roads. At the base of the hill was the first creek bridge. From there, it was about half a mile to the second creek bridge on a built-up gravel road. We would run at full throttle between the bridges.

Christmas morning was always exciting for us. We had a door between the hallway to our bedrooms and the foyer. A right turn would take you to the living room and dining room; going straight ahead would bring you to the family room and kitchen. Then we had a pantry/laundry/mudroom that went to the back door to the garage. There was a chain lock on the hallway door, installed at the top of it to prevent us from unlocking the door and going out before our parents woke. Generally around 6:00 a.m., we would have made enough noise to wake up my parents.

In our family tradition, we exchanged gifts between family members from under the Christmas tree. Santa Claus would come during the night and leave four piles of presents, one for each of the kids. In 1973 Daddy opened the hallway door; we took off like racehorses from a starting gate and nearly collided with three motorcycles. I was given a Honda SL-70, which looked like a small Harley Davidson motorcycle. Jack was given a Honda XR-75 dirt bike. Allen got a Honda 50. We all got brand-new helmets. We were excited to get them to The Place. Now we all could explore together.

The second bridge was about three-quarters of a mile from the top of the hill where our house was located. Jack and I would give Allen about a half-mile head start to the first bridge. Allen would be wide-open on the throttle of his Honda 50, bent down over the handlebars to lower wind resistance. Just as he would hit the first bridge, Jack and I would take off on our motorcycles. At full throttle we would be racing side by side. Allen would be halfway up the hill as his little motorcycle was pushing up against the slope. Every time, just as Allen would look over his shoulder, a few yards from the finish line, WHOOSH, we would come blazing past. I think Jack and I tied most of the time. I don't really remember who would win; we just both knew we had to beat Allen.

The second bridge was constructed of telephone poles and crossbeams. There were two-by-six-inch boards that marked the tire path across the thirty yards of the bridge. There were four-by-four-inch, eight-feet-long wooden poles that marked the edge of the bridge, so you could not drive off the bridge. It crossed over a creek, but the area under the bridge was often flooded, due to the overflow of the Trinity River. One problem was that the two-by-sixes were not pressed together and had space between the slats. The gaps were wide enough that we could get our motorcycle's front tire into the space, and we would have to force it up out of the crack. The boards ended with a crack that created a sharp J turn.

One day, it had been raining all morning and finally stopped. I decided to go for a ride. I got on my motorcycle, put on my helmet, and took off. The road was slightly muddy, and the wood on the bridge was slick. I got my front tire caught in the crack and was trying to get out of it, when I hit the J and it whipped my handlebars around. The back tire spun out. The bike was now on its side, and I was pushing down on the bike as it was sliding toward the edge of the bridge. My front tire hit the four-by-four railing, and I had so much forward momentum that the bike bounced off the four-by-four, and I carried on headfirst off the bridge. I can see it in my mind in slow motion as I was going headfirst and looking at the telephone-sized support pole. Fortunately, the area below the bridge was flooded. I went in headfirst, came up, and grabbed hold of the support pole and shimmied up to the bridge. My adrenaline was flowing. I was mad, standing there soaking wet. I went over, picked up my bike, started it up, turned it around, and went back to the house.

I walked in, soaking wet; everyone stopped and said, "What happened to you?" I had only been gone about ten minutes. I said, "I went off the second bridge." I changed out of my wet clothes and then walked right back outside, got on my motorcycle, and took off again. I was very lucky that I didn't get seriously hurt.

One of the most enjoyable things was to get my mother to drive us down the road in our Chevy Blazer. We would grab our .22 rifles and drive the roads looking for snakes. The land was covered with venomous cottonmouths, or water moccasins, and copperheads, as well as nonvenomous snakes like the diamondback water snake, the Texas rat snake, and garter snakes. We would drive down the road, looking for snakes and snapping turtles sunning on logs or on the banks of creeks and lakes, and give them their demise. My mother absolutely hated snakes. Many of the snakes we killed we would skin. We would dry the skin with rock salt, and after a few days they were ready to hang on the wall in El Lago.

One time, I found a small dead garter snake and put it in the rear pocket of my blue jeans to skin later. I forgot about it, and when the weekend ended, we returned to our home in El Lago. We all piled our dirty clothes in front of the washing machine. The following morning, I had already left for high school when my mother went to put a load of clothes into the washing machine. As she lifted up the pile, my dead garter snake fell out, and my mother went screaming out of the room.

Jack and Allen, who were in intermediate school, were still home. They rushed out, and my mother was screaming, "There is a snake in there!" The two of them waded into the clothes and gently sorted them until they found the dead snake. When I got home from school, mother was still upset. I confessed that it was me and asked her what she would have done if Jack and Allen had not been present. She responded, "I couldn't decide whether to shoot the pile of laundry with a shotgun or just leave for the whole day."

Another time, a few years earlier, I was outside the El Lago house, and some friends came over with a plastic snake. It looked very real. I took the snake and wrapped it around my arm, with the head pinched between my fingers. I put my hand behind by back and walked into the house. My mother was on the phone in the kitchen. I walked in and said, "Mama look," bringing out my arm. My mother screamed, threw the telephone handset into the air, and took off running. The handset hit the kitchen floor, and I look down and could hear Daddy's voice coming out of it, "Joan? Joan? Joan?" I picked up the handset, and now my mother was speaking from the telephone in the master bedroom, upset, telling my father what I had done. It was a bad idea; I paid for it. I never did it again.

Another time, I was hunting along a creek. I saw a big cottonmouth sunning on the bank. I found a stick and pinned its head. I reached down and picked him up. He was big. He wrapped himself around my left arm and was using his body to attempt to pull his head through my fingers. I knew that if I let that happen, I was going to get bitten. We were in a death struggle. I finally choked it to death. In hindsight, I should have used my other hand to grab the snake by its tail and unwrap it from my arm, thus eliminating its ability to use any leverage.

Once, I was with my brothers, and we were walking along a creek, looking for big bullfrogs. The water in the creek was low with a steep bank. We spotted a frog, and I shot it with my .410 shotgun. It started to fall into the creek. Everyone was yelling, "Get it! Get it!" I was trying to get down the steep bank, when suddenly I slipped and slid all the way into the creek. I just held my shotgun over my head, as the rest of my body got completely submerged. I laid my shotgun on the bank and was ducking under the murky water, trying to find the dead frog.

From underwater, I heard, "Whoom. Whoom." I came up and saw my brother Allen standing about ten feet down the bank, staring into the creek. He announced, "It was a cottonmouth, but I got him." I decided to abandon my frog search at that point. True hunters always feel bad when they can't find the game they shoot.

Other times, the lowlands were flooded by the Trinity River. It was after dinner and dark outside. We decided to go frog hunting. One thing about frog hunting is that the frog is already in a position to explode into a leap. In frog hunting, you want to quickly blow the frog's head off before the leap response can kick in. We would take hollow-point bullets and use a knife to cut an X in the top. This created a dumdum bullet that exploded on contact. There was a whole group of us going on this particular hunt. My brother Jack and I were sitting on the hood of the Blazer, with my father, Jim McQueen, Bob Jamison, and the younger kids inside it, and three people were riding on the lowered tailgate. Jack had the best eyes and was the spotter.

A Q-Beam spotlight would help you figure out what species you had found. Spiders' eyes showed up red in the light, and frogs' eyes reflected white. When he would spot a frog, Jack would bang on the hood, and my father would stop. Since it was flooded, one of the only pieces of high ground was the road. Everyone took turns shooting frogs. Jim McQueen was going to show us a trick to use a fly rod to dangle a fishing fly in front of a frog. The frog would lick it down like a real fly, and then we could reel him in. He even bounced the fly off the frog's nose, and it didn't budge. We all had a good laugh at him.

Once you shot a frog, the next step was to hold him by his legs and wallop his head against the rear bumper to make sure it was dead. A good hunter does not want an animal to suffer. At one point, Jack spotted a frog about three feet from the bank, on a large lily pad. We didn't want to shoot it since it might fall into the water. I was sent to catch it. The way to catch a frog is by shining a flashlight at it and moving in to grab him right in the middle of his spine. I was moving closer and closer and had to step into the water, which required slow movement so as not to create any ripples that might spook the frog. I was focusing on the frog, my hand about six inches away; just as I was getting ready to make my quick grab, I saw coming into the top of my lighted circle a cottonmouth going after the same frog. I jumped back, making a big splash. The frog leapt over the snake and got away. Neither the snake nor I was happy, but it was a lucky day for the frog.

We got twenty-three frogs that night. We called it the Night of the Big Frog Kill. We came back and cleaned them, and my mother cooked frog legs for dinner the next night. Bob Jamison showed us a quick way to skin a frog. He took a sharp knife and made a slit at the bend in the frog's back, and then he used water pump pliers to jerk the skin off. Then, cutting through the back, all the innards would fall out. It was quick and fast. I later had to teach my cousins in Mississippi the trick. Daddy would later say, "I loved going into NASA on Monday morning and seeing all the other astronauts with their Corvettes, and I had my Blazer covered in mud and frog blood."

On another trip, we learned from Bob Jamison how to hunt robins. They are a migratory bird, and The Place was located on the robin flyway. We literally had thousands upon thousands of birds that would descend on The Place and the surrounding land. Robins taste just like doves. You clean them the same way as a dove. You start by breaking off the wings, and then you pull the breast meat out. We would stand out in the evenings, next to the road, and wing shoot until we had enough for dinner.

At The Place, we didn't have a TV, so we had to find our own entertainment. Many nights, it was just driving up and down the road with a Q-Beam.

I was fourteen, and my mother let me drive the Blazer up and down the road. At bedtime, we would leave our bedroom light on in the boys' dorm. We had two sets of bunk beds and a desk. We had three tree frogs who lived on the side of our window over the desk: Beauford, Gernie (short for Gertrude), and Fred. We would lie in our bed and watch them attack the mosquitos that were attracted to the light shining out of our bedroom.

We would drive to The Place most Friday afternoons after school was out. Saturday was one of my only chances to sleep in. Every Saturday morning, starting at dawn, we had this crow that would begin to caw from one of the big trees down the hill from the house. He always woke me up. I would try to sneak out the front door to get a shot at that pest. I also tried easing out the back door and sliding around the house to get a shot. No matter what I attempted, that darn crow would always spot me and fly off before I could get him. He was my nemesis.

Bob Jamison also owned a Citabria airplane. The Citabria is a light, single-engine, two-seat airplane with a fixed landing gear and a taildragger. My father would fly from the back seat, while the passenger rode in the front seat and tried to not touch the stick or the rudders. It was designed for aerobatic maneuvers. The name spelled backward is "airbatic."

Bob had a hangar and dirt landing strip next to a soybean field. Every so often, Daddy would take us up flying in it. One time, he had my brother sitting in the front seat, and I crawled in the far back of the plane, where there was a little area for storing luggage. There was enough room for me to ball up and hold on to the seat in front of me and rest on the balls of my feet. We got airborne, and Daddy took us through doing a parabola, just like NASA does in the Vomit Comet. The plane went into a dive and picked up speed. Then my father pulled back on the stick to begin a climb, and then he pushed the nose over. The force continued to push us up, and we began to float in a weightless state for a few seconds. Then the process started over again. I remember the first time he did it; I was in the far back and not strapped in. I just floated in the air. I also remember how much dirt came off the floor and floated around the cabin. It was one of the coolest things I had ever done. We did several parabolas, and it was a blast. Neither my brother nor I was ready to land when we finally did. At one point, my father held the record for doing the most parabolas in NASA's Vomit Comet.

At times, Daddy would let us fly the Citabria. I would put my hands on the stick, and I had to sit forward in the seat to get my feet onto the rudder pedals. I would climb and dive and turn right and left. There is an instrument called the two-minute-turn indicator. It is a tube with a black ball floating inside it. There are two black lines on the tube, and if you can keep the ball floating between the two lines, that would mark a perfect 360-degree turn. My father would say that Joe Engle was the best stick-and-rudder pilot he had ever seen. Joe flew the x-15 and was assigned to be the lunar module pilot of *Apollo 17*. When it was announced that *Apollo 17* was going to be the last lunar mission, Joe was replaced by Harrison Schmitt. Joe later flew on two shuttle missions. My father said that when he flew with Joe, Joe could do all kinds of maneuvers, and the ball never moved. That intrigued my father, who at first assumed the ball was stuck. However, when my father took the stick, he saw that the ball could move just fine. Joe Engel was just that good with the stick and rudder. That story inspired me as a boy, so when I banked the Citabria, I tried to be just like Joe. Unfortunately, the ball moved all over the place.

My father told us a story about his time in flight school. Apparently, a pilot called the tower and said, cocksurely, "Ace on base, got the gear down and locked, and the program knocked." The tower control smugly put the pilot back into place and said, "Go around, Ace on base." We would laugh and use the pilot's expression when bringing in the Citabria to land.

One day, I was just riding my motorcycle down the road when I spotted a narrow cattle path leading off into the woods. I decided to follow it. Deeper and deeper it went into the woods, until I hit upon a large lake. It was beautiful and the perfect duck-hunting lake. Surrounded by trees, its calm water was away from any civilization. I came back and told everyone about the discovery.

My father took us up in Bob Jamison's Citabria the next day. I guided him over the lake. It made a large curve, and from the air we could tell it used to be part of the flow of the Trinity River. But the river had changed course, and now it was just an isolated lake surrounded by trees. We then took the Blazer through the woods to make sure we could get the truck down to it.

It was in the fall, and duck season had started. My brothers and I decided to go duck hunting by the lake. We spread out along the backside of the lake, a few hundred yards apart. I saw a wood duck swoop in, and I shot it. It fell into the water about thirty yards in front of me. I had on my waders, and I started out into the lake. The water was only about two feet deep, but the bottom of the lake was deep silt. With each step, I was sinking deeper and deeper into the silt. I sunk so deep that the water was coming over the tops of my waders. The weight of the water was pulling me forward. Every time I tried to put my hands down to push myself back up, my arms would disappear in the silt. I could see myself drowning in two feet of water when my waders completely flooded and pulled me down into the silt. I was getting panicky. We had a rule that if you were in trouble, you were to fire your shotgun rapidly three times, reload, and do it again. I did that. No one came.

I started just firing off all my shotgun shells as quickly as I could. My brothers decided that I was making too much noise for them to hunt, and they decided to see what was going on. I was so relieved to see them. They managed to find a long, dead tree and push it out me, and I used it to pull myself free. We learned then never to go into the lake again. On our future duck hunts, we smartened up and used a Labrador retriever named Co-Pilot to fetch the ducks.

With so many lakes on the land, we had many areas to hunt and fish. We had one particular lake that had lots of bass in it. I used to take my motor-cycle and fishing rod and tackle box and ride down to the lake. I would generally use a plastic worm with the hook turned back into the worm. This allowed the worm to be dragged across the bottom and not get snagged on logs or other plants.

The downside to this type of fishing is that when a bass hits on the bait, it doesn't swallow it right away. It generally holds it at the edge of its mouth. If you try to set the hook too early, it just spits it out. It took patience to watch the line move around the lake. Finally, after what seemed like an eternity, I would try to set the hook. Most of the time it worked. Bass are a great fighting fish, and it was amazing to see them splash out of the water. I can remember I was going to try to cast around a large tree that was about five feet to my right. I attempted to fling my lure, and it hit the tree and fell straight down into the water. Wham! A bass hit my faulty throw, and we were off to the races. It was big. I would bring home my catch and clean the fish, and my mother would cook them for dinner.

We would go out in the afternoon and set crab and crawfish traps. Over the course of an afternoon, we would catch a trash can full of them. We then had huge crab and crawdad boils, relaxing and eating for hours.

One time while at The Place, we decided we needed to clear some land for a garden. Bob Jamison arranged for us to borrow a tractor with a disc harrow and a Bush Hog. The Bush Hog is like a huge lawn mower towed behind a tractor that can chop up bushes, grass, and very small trees; it is great for clearing the land. A disc harrow has concave, scalloped metal discs that are set off at oblique angles. It rips through the soil, cutting up the ground into small chunks so that it is ready for planting. The disc we borrowed had about six bars with five discs on each bar.

The farm from which we borrowed the equipment was about five miles from The Place. The plan was to place the disc on top of the Bush Hog and attach the Bush Hog to the tractor. I was then able to lift the Bush Hog and transport them back to The Place. I had grown up driving tractors on my cousins' farms in Mississippi. I loved it. I had been trained to push down on the clutch and brakes at the same time so as to not stall out the engine when trying to stop. Of course, the clutch was also used when one needed to shift gears. The tractor engine throttle is on a lever located next to the steering wheel. This allows the farmer to set the speed without keeping his foot on the accelerator.

The farmhouse was located on the top of a hill. The two-lane road that ran in front of the farmhouse descended the hill to a bridge that crossed over a creek, before heading back up another steep hill. My father and siblings were going to follow behind me with the flashers going on the Blazer to warn other drivers about the tractor being on the road.

I got the tractor started, lifted up the Bush Hog and disc, and started down the gravel driveway to the road. I shifted the tractor into high gear as I got ready to make the right turn onto the road. When I looked behind me, Daddy was following, so I added the throttle level and swung onto the road.

As I crossed the cattle grid, I was immediately in trouble. A cattle grid has spaced-out pipes that prevent livestock from crossing out of a property. The pipes caused the tractor and Bush Hog to bounce. The Bush Hog and disc made the tractor very tail-heavy, and the front tires were bouncing and barely touching the road. I was heading down this very steep hill and swaying in the road. I started to turn the wheel to adjust, but each time, I overshot the arc. I stood on the clutch and braked, trying to stop, while continuing the wild oscillations down the hill. I was using up both sides of the road; fortunately, there was no oncoming traffic. The tractor was at almost a forty-five-degree angle as I slid across the bridge and started up the other side. Going up the steep hill bled off my speed, and I was able to pull over to a stop. I was so full of adrenaline that I was shaking.

My father pulled over behind me and came running up to me and gave me a hug. He said, "I thought I was going to watch you get killed."

I said, "I didn't know what to do—it just kept getting worse." Apparently, it is called "p-i-o," for pilot-induced oscillations. The pilot's controls are delayed, so the pilot keeps putting in input. As the machine falls farther and farther behind each correction, it eventually goes out of control. I replied, "That was an intense p-i-o." I was amazed that there was a name for such a thing. My brothers jumped out of the vehicle and ran up yelling, "That was wild!"

We took a few minutes to calm down. We had more hills to go to get back to The Place. We decided that I was going to stay in the lowest gear and not drive too fast, and if things started to go bad again, I would just lower the leveler and drop the Bush Hog into the road to stop.

On the adjacent land, we had many oil wells. We didn't own any of the mineral rights, so we didn't get any revenue. However, that didn't mean we didn't play with them. The pumping jack is a big piece of machinery. It has a long beam held in the middle on a rocker. The rear of the machine is connected to an engine that has arms that move the beam rocking back and forth. Right over the drill hole is a very large metal curved head that holds the cables that rock back and forth and pull the metal rod back and forth in the ground to pump out the oil. At the bottom of the stroke, we would jump on the bracket that held the cables to the rod and then place our hands on the cables and ride up and down on the pumping jack. We would take a piece of two-by-four board and place it in the gap of the cables and the medal head, and it would cut through it like a hot knife through butter. One time, my father got on it to try it. My brother Jack quickly called out for him to lower his hands. With his original grip, he would have lost all this fingers. They were big, they were fun, but it was dangerous.

Another thing we used to do was fill up a gallon of crude oil into an old water jug from the oil well's spigot, and we'd go pour it on a fire ant bed and light it on fire. Seems a bit cruel now, but I hated fire ants. I have been bitten too many times. When I was about six years old, we were visiting Sessums, Mississippi. I was playing in a cow field and got into a fire ant bed. I am convinced that fire ants can communicate with each other. The first one doesn't bite you when it first starts to crawl on you. They wait until you are completely covered, and then they send out a signal for all to bite at once. I was covered from the top of my head to my feet when they started to bite. I ran scream- ing and crying into my grandmother's house. She immediately stripped off my clothes and put me into a tub of water, along with some vinegar to reduce the pain. I had ant bites all over my body. For the rest of my life, I have been at war with fire ants.

We had lots of pine trees at The Place. Since we all loved to hunt, it was only right that we "shoot our Christmas tree." We would drive around in the woods and look for a pine tree about thirty feet tall with a nice top. We would aim about ten feet down and attempt to shoot the top of the tree off with one shot from the shotgun. It was extremely gratifying to be able to do it with one shot. Some people might take two or three shots.

Generally, Daddy would invite another astronaut family out for the fun and a chance to take home a fresh Christmas tree. Unbeknownst to the astro- naut victim, Bob Jamison would have the local game warden show up at the right time and ask what was going on. Bob or our father would tell the warden that we were hunting Christmas trees. The game warden would turn to the invited astronaut and say, "Let me see your Christmas tree hunting license." The astronaut would stammer, "What?" Daddy and Bob would break out these pieces of paper, saying, "Here are our licenses. Even the ones for the kids." They would carry on, "Of course, everyone needs a hunting license." The flabbergasted visiting astronaut would have to admit that he didn't know he needed a license, to which the game warden would say, "Well, I am going to have to write you a citation and confiscate your shotgun." They would let this farce play out for a while, and then they would have a big laugh and let the poor victim in on the joke.

Some weekends, we would be at The Place when our father was coming home from the Cape. He would buzz the house and put on an airshow for us. I can remember the high-pitched whinnying of the engines of the T-38 as he cleared the treetops. One day, Daddy came all excited into the house. He was exclaim- ing, "I made Fred duck! I made Fred duck!" My mother was like, "What hap- pened?" Fred Fatherly was a crop duster and shared my father's love of flying. Some days, we would go up in his crop duster and fly around. If Fred had any leftover insecticide from doing a job, he would fly over The Place and drop it to help keep the mosquito population down.

On this particular day, our father was flying back and saw that Fred was landing his airplane on his dirt strip. Daddy said, "I flew down low and headed straight toward Fred." Fred had his back turned, so when he suddenly heard the roar, he turned and ducked as my father flew over him.

My mother coolly said, "Well how low were you?"

Daddy replied, "About five feet off the deck."

My mother then asked, "How tall is Fred?"

"About six feet," was his response.

My mother stated, "I guess it is a good thing Fred ducked, or you would have taken his head off." She quickly put a damper on his excitement.

With my father gone off training much of the time, my mother had to play both parental roles. She was the disciplinarian. She would say, "I don't want to have your father come home and be the disciplinarian. First, he is gone too much to make the punishment fit the crime. Second, I don't want that every time you first see your father, he is having to spank you."

When I was fourteen, my mother and I were going through a rough patch. I was a rebellious teenager. She was frustrated, and I was angry. I don't remember the argument, but my mother said, "If you don't like it, you can leave!" I responded, "I am leaving!" And my mother replied, "Fine, go!" I turned and walked out the front door.

I didn't have a plan. My father was at his office at the space center. So I decided to walk to NASA and say goodbye to Daddy before I left on my running-away strategy. It was about three miles from our house to the NASA complex. I hiked to the front of our subdivision and then along Highway 1 to the front gate of NASA. I talked my way past the gate guard and then to Building 4. The astronauts' offices were on the third floor. I got off the elevator and walked to his office. I walked up to his secretary and asked if my father was around. She replied, "He is off training. Can I take a message?"

I replied, "Yes, tell him I just came by to say goodbye. I am leaving."

She looked aghast and said, "Hold on," and picked up the telephone.

I was like, "I need to get moving—it is getting late in the afternoon." I didn't have much money on me, no backpack or sleeping bag, not even a change of clothes. I just knew that I wanted to leave. The secretary got Daddy on the phone, and I spoke to him briefly. I told him about the fight and that I was leaving and just coming by to say goodbye.

He replied, "Hold on a second; I will be right over."

I was like, "I need to get moving."

He said, "Please wait. I will be right there." He came over and picked me up and said, "I got to finish up some work, and then we will talk about it." We got into our car and drove over to a training building. We walked inside and into a dark room. There was a light table with images of the moon.

I was introduced to Farouk El-Baz, the world's leading lunar geologist. He spoke with a heavy Egyptian accent. Daddy sat me down between them and started going over the photographs. These were photographs that my father had taken on his mission, and he was discussing them with Farouk. During my father's flight, he had taken some spectacular pictures of a large crater on the back side of the moon. The astronauts were not allowed to name anything on the moon. Farouk started to refer to this crater as the SAR crater. Lunar geologists went with it, not understanding the term's meaning. It was my father's initials: Stuart Allen Roosa.

I recall that Daddy pointed to a boulder and I climbed up close to take a look. "That boulder is bigger than our house," he told me. I was like, "Wow." I was there for a couple of hours getting my lunar geology lesson from Farouk.

When the meeting concluded, Daddy said, "Let's head back to the house, and we can discuss it."

"No, I need to go. Mama told me to go," I said.

He countered, "I am sure she is no longer that mad. We can talk about it." I reluctantly agreed, and we made the short drive to the house. I can remember walking back into the house. My mother had calmed down, and Daddy dissipated the situation. I never tried to run away again.

Of course, no recounting of youth would be complete without a tale of youthful love. Mine was Denise. When I was sixteen and a junior in high school, my best friend, Bruce, and I were cutting up in English class. Our English teacher had a nervous breakdown during the year and was out for a period of time. She had recently returned and didn't want any stress. Apparently, our cutting up in class affected her, and she made the decision to move my desk to the opposite side of the room.

The next day, a girl named Denise was transferred into our class. She was assigned to sit next to me. I am sure most schools have a "Denise"—a pretty and somewhat unobtainable head-cheerleader type who is dating the handsome quarterback; or some girl who was a junior dating a senior and, the following year, is now dating a college guy; or the young woman who is just too mature for her cohorts at school.

Denise was beautiful, extremely beautiful. She was dating a twenty-five-year-old firefighter. She was sixteen. She moved through the halls as if we were all ghosts. She was never rude, nor haughty, nor condescending; it was just that no one in the school was in her league.

Denise and I started with small talk. Here I was, one of the littlest guys in the school. I was not going to compete against the firefighter, and I had never dated up to this point in my life. We would talk during the beginning of class and then go our separate ways when the bell rang. As luck would have it, we were assigned to work together on an English project. We traded ideas, and over the course of time, I figured out how to make her laugh.

As we were ending the school year, the junior class picnic was approaching. One day in class, she looked at me and stated, "You are taking me to the picnic."

I was stunned and mumbled, "Sure." The day arrived, and we boarded school buses and went to a large park. There were various activities and picnic tables, food, games for the students to play. At one point, we went for a walk and ended up sitting down among the trees. We were talking, and I remember she was in midsentence when I leaned in and kissed her. It was the bravest thing I had ever done. She kissed me back.

Our first date was a disaster. I had stolen a bottle of rum from my parents' bar and purchased some cokes from the UtoteM near our house. The drinking age in Texas was eighteen. And given older brothers, fake IDs, and friends working in the grocery stores, there was a lot of underage drinking. I wanted to impress Denise.

I picked her up in our nine-passenger station wagon and drove the thirty-five miles to Galveston Beach. It was a beautiful day, and we found a nice place to park and put on our bathing suits to go play in the surf. She was beautiful in her bikini. I was holding her hand as we waded out through the waves. We were in waist-deep water when suddenly my right leg crashed through an old sunken metal fifty-gallon drum. I immediately tried to pull my leg back out. I had scraped most of my leg. I was bleeding, and the salt water was stinging my wounds. We had not been in the water a total of five minutes, and the pain was too intense to stay in. We went back to the car, cleaned up my wounds, had a few drinks, talked, and headed home with my bloody leg. She was nice about the whole situation.

We dated. She taught me a lot. Unfortunately, my father retired from NASA at the end of my junior year, and so we had a short dating period. My friends were all amazed that I got the girl "who was out of everyone's league."

23

The Apollo Groupie Scene

Astronaut families lived in Houston, though the astronauts did a lot of training at Cape Kennedy, Florida. The bar at the Holiday Inn, Coco Beach, was the astronaut hangout spot. It was where the groupies would go. Astronauts were the biggest rock stars on the planet. Why? They were the only ones truly leaving the planet.

My father said Alan Shepard was the most intimating man in the world. Daddy said Al never gave anyone a compliment, except for once. Alan Shepard was a ladies' man. One time at a dinner party, my mother leaned over to Louise Shepard and said, "Those are some nice cuff links Al has on." Louise responded, "Yes, one of his love bunnies gave those to him." Apparently, Al was in Las Vegas and ended up partying with Frank Sinatra. The only compliment my father heard Al give was this: "Frank really knows how to work the women." Daddy would laugh and say, "Man, I would have loved to have been a fly on the wall to witness those egos going at it."

At one point, the astronaut divorce rate exceeded the national average, and NASA hired a psychiatrist to come study the problem. My father said there was no problem. He explained that long before they became astronauts, most of them married their high school sweethearts, and their wives dutifully followed their husbands from base to base, raising the kids, making the household work, and doing the officers' wives' politics at each base. Daddy would say, "A wife might not help you get promoted, but she sure could stop it."

So now these guys were selected for NASA, darlings of Houston society, with groupies everywhere, and suddenly they would look at their wives and decide they were no longer cutting it. Daddy looked at it very differently and would say, "It took my getting to NASA to get Joan back to the level of society she was used to."

I once asked my mother if a particular astronaut had a mistress. My mother told me that he did and his wife had a boyfriend. "Wow!" I thought.

There was another story about a wife knowing her husband had a woman down at the Cape. The wife and mistress were both at a reception at the Cape, and the wife went over and poured a full drink over the woman's head, saying, "I know who you are."

My mother was often asked if she was nervous during my father's flight. She would smile and say, "At least for nine days in a row, I knew where he was and who he was with."

The first astronaut to divorce and marry his mistress was Donn Eisele of *Apollo 7*. This sent shock waves through the Astronaut Wives Club. There was a lot of discussion about whether the new wife should be allowed to participate. My mother was of the opinion that if they were married to an astronaut, you could not exclude them. As more and more astronauts divorced and remarried, my mother created the Original Wives Club. She was particularly irked at one woman who married an Apollo astronaut after his flight and subsequent divorce. When asked whether she was nervous during his flight, the new wife would reply, "No, not at all." My mother would vent, "Of course not, she was not married to him at the time." She should have responded, "We weren't married at the time of the flight." Instead of acting like she was there and wasn't nervous at all.

After each mission, an informal astronaut-pinning party was held. The crew would be presented with their gold astronaut pin to replace their silver one. The point of the party was to "roast" the crew so that they were "brought back down to earth." This was an ego check, so they understood they were not as big of a deal as they thought they were. My mother was always shocked that Jack Swigert was allowed to bring a date to these events. She felt these were very personal and should remain within the astronaut family. This was largely because, through the roast, Jack's dates would learn about the astronauts' personal lives but had no reason to keep their secrets. My father never spoke of what he learned there, and my mother considered it all "top secret."

24

Apollo 17

NASA had announced that the Apollo lunar program would end with *Apollo 17*. The original plan had been for the Apollo program to end with *Apollo 20*. Due to my father's rotation, it was very possible that he would have been the commander of *Apollo 20* and thus the last man to walk on moon. Instead, because of the early cancellation, he was moved to the backup crews of both *Apollo 16* and *17*. Since he was already trained in the command module, it only made sense for my father to remain in that role.

My father was training down at the Cape and was in the office when someone came in announcing that they had just seen a rattlesnake in the lunar surface training area. This was an area designed to look like the moon and allowed the crews to train on the lunar rover, a four-wheel drive dune buggy. My father asked where the snake was located on the track. He jumped into a car and drove to the lunar training area. He said, "You could hear the rattlesnake rattlers rattling," even before he turned off the car engine. He got out and found the snake. It was a huge rattlesnake, all coiled up and ready to strike.

My father had been taught how to kill a snake with a rock, particularly a flat, circular rock. A big rock does not inflict a fatal blow. A sharp flat rock thrown hard at the snake will break its back and kill him. After a couple of throws, he had killed the snake, and he then used a sharp rock to cut off its head. Removing the snake's head is necessary because, even dead, the snake's fangs can still inject venom. So he removed the threat by burying the snake's head on the track. He then took the snake's body back to the car and put it in the trunk. He said that, lifting it as high as he could, the snake's tail was still dragging on the ground. He figured the snake was over seven feet in length.

He drove back and put the snake's body into a trash bag. He then sneaked into the offices and coiled it up underneath of the desk of Gene Cernan, the commander of *Apollo 17*. My father took the time to tuck the cut neck under the body so no one would realize the head was missing. He then had one of the secretaries tell Gene that he had a telephone call. He and a couple of other astronauts were sitting at their desks "working away," as Gene came in to take the call. Apparently, when Gene went to sit down at his desk, his foot hit the snake; he looked under the desk, screamed, and pushed the desk chair so hard that it made a dent in the wall. The phone flew into the air, and Gene ran out of the room. Daddy would laugh and laugh every time he told that story. He would say, "That dent is still in the wall."

In the fall of 1972, a few months before the launch of *Apollo 17*, the CIA intercepted a message from a Palestinian terrorist group called Black September stating that they intended to kidnap some of the Apollo astronaut families. This group was responsible for the assassination of the Jordanian prime minister Wasfi Tal and for the Munich massacre, their most publicized event, in which eleven Israeli Olympic athletes and officials and a West German police officer were kidnapped and killed during the 1972 Summer Olympics in Munich, Germany.

Because of that threat, we were given Secret Service protection for several months. We had an agent who parked in our driveway twenty-four hours a day. My father thought it was strange just to be parked in our driveway, when behind our fence in the backyard was a forest area that would have provided covered access to and from the house.

Our school bus stop was in front of our house. The bus driver would often ask, "Who's the guy sitting in the car?" We just ignored the question, as if we didn't know what she was talking about. About three weeks before the launch, scheduled for 7 December 1972, we were moved to Patrick AFB in Florida. The families of both the prime and the backup crews were brought to the base, and we all lived on the second floor of our two-story building.

We had a guard desk on the first floor to limit access, and we had a guard sitting at the top of the steps, with a machine gun lying on the desk. The Secret Service had two large Winnebagos full of agents. We even had a telephone in the apartment that rang only in the command post inside one of the Winnebagos. Any time we wanted to go outside, we had to have two agents for each of us.

One time, I got cabin fever, and my brother Jack and I wanted to go outside and throw a football. We had four agents watching us, and we were standing in the middle of Patrick AFB. There was another time when my mother needed to go to the PX, the base's commissary, to purchase a sewing kit. The lead agent suggested that instead of two agents taking my mother, he would just send one to the commissary who could bring back the sewing kit. My mother agreed.

At some point, the agents must have felt sorry for us. They announced that they were going to take us to Disney World. We were all very excited. In those days, Disney World still had these ticket books with tickets from A to E. The A tickets were for the smallest or least popular attractions, and the E tickets were for the most popular. The books had a limited number of tickets for each level.

When we arrived, we were asked what ride we wanted to try first. I replied quickly, "The Haunted Mansion!" It was a famous ride. Our security detail walked us to the front of the line, bypassing who knows how many people who had waited who knows how long.

The Haunted Mansion was amazing. At one point in the ride, a ghost appeared in the car next to you. We got off all excited and said, "Can we do that again?" Because we had a security threat against us, back to the front of the line we went. We had a great day. After receiving the VIP treatment, I decided I never wanted to go back to Disney World until I could do it again like that.

With the families still under the terrorist threat, the crews went into preflight quarantine for three weeks. We got to go speak to Daddy through the glass windows.

The families had set Secret Service details, and through the other crewmembers' families, my father had been told that the best-looking agent was assigned to my mother. The astronauts would razz each other about that.

At one point, my mother was at a prelaunch party on the beach. A very large carnival tent had been erected. My mother arrived in the crowded space, and an agent asked to hang up my mother's coat. A second agent went to get my mother a glass of wine. While my mother was standing alone, our close family friend, the world's leading lunar geologist and an Egyptian, Farouk, came and grabbed my mother's hand, saying, "Joan, you have never met my wife—come with me." He started pulling my mother through the crowd. My mother was yelling, "Farouk, Farouk, stop, stop!" Farouk did not hear over the noise of the party. He pulled her out the rear of the tent onto the beach, and my mother said, "Farouk! Stop! You don't understand."

Meanwhile, the two agents returned to where they had left my mother inside and were told that some "Middle Eastern guy" grabbed her and took her that way, pointing to the rear of the tent. "This is it," they must have thought, "the kidnapping is happening!" As my mother was returning toward the tent, the two agents were running over people, pushing them out of the way, with their hands inside their jackets on their pistols. When they came out of the tent, my mother was already yelling at them, "Everything is okay. Everything is okay." They were not happy, "Joan, you can't do that to us." She apologized and explained the situation.

Finally, we made it to launch day and went to the astronaut viewing area to watch the liftoff. That morning, my father had been released from preflight quarantine. We got into our nine-passenger station wagon, and we watched as all the families and guests boarded the NASA buses to receive a police escort to the astronaut viewing area. We were stuck in traffic caused by the police clearing the way for the convoy. Daddy pulled in between the last two buses. There were about six buses in the convoy.

The bus driver behind us started honking his horn. Daddy was laughing, my mother was telling my father to be careful, and we kids were just taking it in. It was a short drive to the NASA property. Daddy pulled out of the convoy, passed buses, and wound in between them. People on the buses would point at the car and wave, recognizing our family. We would all wave back. He would then pass another bus, get honked at, and laugh even louder.

Now on the Kennedy Space Center property, we then passed the first bus, and my father ducked in behind the police car. The police officer threw on his brakes and pointed for my father to get out of the convoy. Though Daddy hit his brakes hard, we almost rear-ended the police car as he waved us out from behind him. So he quickly pulled out and then sped to the viewing area. We were out of the car as the buses pulled in, and everyone was laughing at how my father beat the police escort. He had a great laugh on that day.

Apollo 17 was the one and only Apollo launch to be scheduled at night. The bright spotlights lit up the massive rocket. Like a modern-day pyramid, it was a technical marvel of its time. "Three, two, one. We have liftoff." The first stage lit up the night. It was suddenly as bright as daylight. The sound, the ground shaking, the intensity of the flames, and then it was over. The Apollo program was finished. The most powerful rocket ever created, silenced for eternity.

After the launch of *Apollo 17*, we spent a few more days at the Cape. Daddy took us to the astronaut living quarters. He showed us his room, the dining room, and the gym.

While at the Cape, we got to meet Werner von Braun. Over the course of my father's career, he met thousands of people. But the most impressive man he said he ever met was Wernher von Braun. Here was a guy who at twenty-six years old stood face-to-face with Hitler. My father said, "You could be standing in a reception room with your back to the door, and you would instantly feel von Braun's presence when he entered." My mother would say, "Every time I met von Braun, he was so German, he would click his heels together and give me a slight bow." I got to meet him on several occasions. He was very kind to me.

In World War II, Wernher von Braun trained himself on rockets by creating the v-1 and v-2 rockets that would be fired at England. In the closing days of the war, Hitler ordered the s s to go to Peenemunde and kill all the scientists and engineers. This word got leaked to von Braun, who then told the other scientists and staff of the facility. The Russian lines were closer, so many defected to the Russians. They became the basis of the Russian space program.

Von Braun wanted to defect to the United States, but the U.S. lines were farther away. As he and some fellow scientists attempted to drive to the U.S. lines, they ended up in a car wreck. With his arm in a brace, he surrendered to a U.S. Army private riding a bicycle. A U.S. Army intel captain realized who von Braun was and quickly flew him to the United States.

With the advent of Sputnik, the first satellite, the Soviet cosmonaut Yuri Gagarin was the first man into space, and the United States was behind in the space race. Many test rockets blew up on the pad or shortly after liftoff. There was a decision to bring the Germans into high gear to help NASA. So NASA made von Braun the chief architect of the Saturn V program.

It was after Alan Shepard's Mercury flight that President Kennedy made the famous speech that the United States "should commit itself to achieving the goal, before this decade is out, of landing a man on the moon and returning him safely to the earth." Daddy admired Kennedy's boldness. My father always reminded people, "The United States only had a total of fifteen minutes of space time, and President Kennedy committed us to going to the moon. We walked out at night and looked up at the moon, and we knew our mission objective."

25

Reflections of an Apollo Command Module Pilot

My father was often asked to give speeches or to be interviewed on his experiences in the Apollo program. Often during a speech or a press event, he would be asked, "How do you feel getting so close to the moon and not walking on it?" He would reply, "We are all fighter pilots. We like to fly alone. While in the history of mankind there have been twelve men who have walked on its lunar surface, there have only been six who have soloed around it. That puts us in a smaller club."

He would often end his speeches by saying, "Go out and look at the moon tonight. There are six American flags flying on its surface. No other nation can make that claim."

He was also frequently asked about the top three experiences he could relate from his mission. His reply was always as follows. First, he explained that the liftoff was much more violent that anything they had experienced in the simulator. The rocket shook from side to side. Daddy was not one to exaggerate, so I know it must have been bad.

He would also talk of being at lunar distance and looking back at Earth. He would say, "Space is blacker than black. The pictures do not capture the true blackness of space. You would look back at the earth and see the white of the clouds and the blue of the oceans; it was hard to make out the continents. You would think, 'That is everything I know: grass, green trees, where I went to school, my family,' and you can reach up and cover it all up in the palm of your hand. You knew you were a long way from home."

Lastly, he would talk about growing up as a poor kid in Oklahoma. His family raised chicken eggs to sell to the grocery store, had only an outhouse, and bathed in the washing tub on the kitchen floor. Here he was on the backside of the moon, alone, and unable to talk with mission control. He would look down at the lunar surface on the far side of the moon and say, "How did a poor country boy from Oklahoma ever get to be here?" He was only the third person in the history of mankind to be in that position. The American dream came true for him.

As the command module pilot, he was assigned to name the command module. I can remember that we were all sitting in the backyard of my cousins' house in Sessums, Mississippi, and we were throwing out potential names to my father. We were throwing out air force themes, space themes, country western ideas, and really anything that came to mind. Daddy thought about it and decided he wanted to honor the U.S. Air Force and specifically our first pilots—the Wright brothers. He named his capsule *Kitty Hawk* from Kitty Hawk, North Carolina, where the Wright brothers made the historic first flight of an airplane.

I watched my father give many speeches, as did all the other Apollo astronauts. They generally all had the same reply to one particular question: How do you go to the bathroom in space? Each would reply, "With difficulty. Next question."

At times, my father might relax before a crowd, and when the question came up, he would tell the details and have everyone laughing. He would start by explaining a urine dump. Each crewmember had a plastic sheath that attached to the urine tube. Once voided, the crewmember would flip a handle and expend the urine overboard.

In the freezing cold of space, the urine would immediately turn into snowflakes. Daddy would say, "Urine made for prettier snowflakes than dumping regular water. The yellow color would make it really stand out, and they were big, beautiful snowflakes."

If he told that story, then invariably, he would get asked, "How about doing a number two?" He would remind everyone that the inside of the command module was about the size of an SUV. They were pretty close quarters. So he would launch into the story:

> On the way to the moon, you have some room for privacy. Because you have the command module and the lunar module attached to each other. The privacy goes away on the return trip since the lunar module is used up.
>
> So when you are coming back from the moon, you are being pulled by gravity, and the trip takes about three days. If someone had to do a

number two, it was generally the entertainment of the day. Two guys would move to the side of the capsule to allow the most room for the participant.

The first thing you had to do was get completely naked. Next, you had a big clear plastic bag, the size of a trash bag, that contained Velcro straps attached to the opening. There was a strap that went around your waist and then two straps that went around your legs; consequently, the bag was now strapped around your butt.

Then you would do your thing. The problem is that in zero gravity there is nothing to make "the mass" move to the bottom of the bag. You would wiggle, do what you could, to get it to float away from your body.

Then you had to carefully undo the Velcro straps, reaching around to close up the top, without letting anything escape. You would then seal the bag, and inside of the bag was another small plastic bag that contained a disinfectant. You had to break that bag and then knead the chemical into "the matter." Once that was done, you floated across the capsule and put it into a special storage area for the scientist to study later to understand your intestinal workings.

As he would walk around the stage, demonstrating these actions, he would have the crowd laughing. He would generally close with the story that fellow crewmember Ed Mitchell went eight days without a bowel movement. However, he was passing very noxious flatulence. Finally on day eight, he would say, "Al ordered Ed to take a dump."

26

The Last Flight of Apollo

By 1975 the U.S. space program had lost its wonder for many Americans; they were distracted by the recent events of the Nixon White House, by the war in Vietnam, and by massive, ongoing, and generational social change. The *Apollo 17* flight would be the last mission to the moon; it would be followed by the Apollo-Soyuz launch. Even so, we were acutely aware that this was the end of an era—the Apollo era—and though many astronauts would remain with the space program as it moved forward, many others would call it a day with the last of the Apollo flights.

My mother and father were always mindful of the historical dimension of these occasions and brought us to this final launch with that in mind.

It was 15 July 1975, and we were there for the launch of Apollo-Soyuz. As we had before, we were standing in the astronaut viewing area at Cape Kennedy, Florida. Since early in the Apollo program, NASA had cleared out a patch of land as a viewing area and placed bleachers on the grounds, which were surrounded by marsh. They would put a rope around the perimeter and build some wooden outhouses every time there was a launch, and afterward, NASA would clean up the patch of land. The purpose was to have an area where astronaut families and friends could watch the launch without media attention. If something were to go horribly wrong, the family's immediate grief wouldn't be broadcast worldwide.

It was the first time I was going to see the launch of a Saturn IB. It looked about half the size of the massive Saturn V rocket. NASA had built a stand to raise the Saturn IB command module up to the crew gantry way. The whole rocket was made of spare parts from the Apollo program. I had taken an interest in photography, and my father had borrowed a Hasselblad camera from NASA to photograph the launch. I was busy setting up the camera on the tripod, when I heard Daddy's voice calling my name.

To be honest, at first, I ignored it and kept setting up the camera. But he persisted. So I turned and started walking toward a crowd. My father was booming, "I have it under control. I have it under control." I walked up, and Daddy said, "There is a snake in the outhouse. Go get it out!"

I was taken aback and replied, "Is it poisonous?"

"Don't know," he replied.

I happened to see an elderly gentleman standing with a cane, so I asked him if I could borrow it. I walked over to the door of the outhouse and opened it toward the edge of the clearing. I looked inside and could see a large black snake coiled up in the corner. My adrenaline started pumping. My first thought was that it didn't look fat enough to be a water moccasin and that it wasn't the right color to be a rattlesnake, though both are abundant in the marshes on Cape Kennedy.

I used the cane to pin the snake's head to the floor and reached down grasping it right behind its head. I lifted it up; it was over five feet in length. It was longer than I was tall. I determined that it was a nonvenomous king snake. I stepped out of the outhouse, and the crowd gasped at the size of the snake. I took a few short steps to the edge of the viewing area, wound up, and flung the snake as far as I could into the marsh. Everyone started clapping, and Daddy started saying, "Thank you. Thank you. I told you I had it under control." I gave the man back his cane and said, "Thank you." He laughed and said, "Anytime."

My father, still absorbing the accolades, shook my hand. I walked back down to where I had the camera set up and waited for the launch. The Saturn 1B lifted off with the crew of Tom Stafford, Vance Brand, and Deke Slayton at 3:50 p.m. The mission was to conduct a handshake with the Soviet crew of Alexei Leonov and Valeri Kubasov. With that liftoff, the Apollo program ended.

27

Leaving NASA

In early 1976 my father decided to retire from the U.S. Air Force and leave NASA. He told us that the space shuttle program was behind schedule. He was unsure of when he might fly into space again, and flying the shuttle would not be anything like Apollo.

It was time to get on with his life. He interviewed with and decided to go to work for a company called U.S. Industries. He would be their director of business affairs for the Middle East. At that time, the Middle East was busting out with oil money, and every company wanted their share of it. Originally, the plan was for our family to move to Beirut, Lebanon. However, the Lebanese civil war broke out, and my father had to choose between Cairo, Egypt, and Athens, Greece. He chose Athens because the communication back to the States was much better.

I recall that one day, Daddy called me in and sat me down. He said, "You are at the end of your junior year. Next year, you will be the big man on campus, a senior. How do you feel about us moving away from all your friends?"

I replied, "We are going to Europe, right?"

"Yes," he responded.

"Let's move," I countered.

We hugged.

In June 1976 my mother took us to Ellington AFB, and we sat on the flight line and watched my father do touch-and-goes with his NASA T-38. After about thirty minutes, he taxied over and stopped in front of base ops.

This was his final flight of his active-duty career. He got out of the jet, unstrapped his helmet, and walked toward us, and we walked toward him.

My mother gave him a kiss, and we all hugged him. As we got into the car, Daddy said, "That's it."

Shortly thereafter, we boarded an Air France fight to Paris and began the first leg of our move to Europe. My days of growing up with NASA were at an end. When the plane took off, I looked out the window and wondered what new adventures life would bring. With that, the Apollo chapter was over.

28

My Father's Passing

On the Monday evening of 21 November 1994, my parents and sister arrived to celebrate Thanksgiving at my apartment in Arlington, Virginia. The weather was warmer than average; it was sixty degrees.

My father had planned for us to go to the very swank Inn at Little Washington, located in a rural area outside the DC metro area, the following night and arranged for a limousine to take the four of us there. The Inn at Little Washington is always ranked as one of the top restaurants in North America. We drank champagne in the back of the limo as we made the one-hour trek from my apartment to the Inn. All the critics rave about the food at Little Washington, with its molecular gastronomy. I remember one small dish that suddenly vaporized in my mouth; it tasted amazing.

After dinner, we made our way home, and Daddy and I sat out on my balcony drinking a beer. My apartment had a view looking up the National Mall and a clear view of the Lincoln Memorial, the Washington Monument, and the Capitol.

At some point, he got up and spat up a little blood. I went to get a rag and told my mother and sister, who were watching TV. They both walked out to check on him, and my sister told me he had done the same thing on the drive up to DC.

I am fortunate that I rarely get sick. In hindsight, this may not have been a good thing, since I didn't have a regular doctor. I had seen a doctor for a minor ailment a month before and made arrangements for my father to see him on the next Wednesday afternoon. They did a blood check and admitted him for acute pancreatitis to Northern Virginia Community Hospital, a small, local hospital.

I soon became an expert on acute pancreatitis, a sudden inflammation of the pancreas. The pancreas is the large gland located in the upper part of the abdomen, behind the stomach. It produces digestive enzymes and hormones. In pancreatitis, enzymes that normally are released into the digestive tract begin to damage the pancreas itself. The gland becomes swollen and inflamed. More enzymes are released into the surrounding tissues and bloodstream. Essentially, your digestive enzymes start attacking your internal organs.

The next day was Thanksgiving. My mother cooked a meal, but no one was really in the mood. We visited Daddy in the hospital, and he looked weaker. Each day, we had a different doctor, since it was the long holiday weekend. I remember on Sunday, I put him into a wheelchair and rolled him into a shower with a stool. He so enjoyed getting to clean himself, and then I helped dress him.

I called my brothers, Jack and Allen, to inform them of the situation. It was dire. Allen said it hit him hard when I told him to "pack a suit in case we have a funeral." Jack was going through the U.S. Air Force Weapons School. They both flew in on the Friday after Thanksgiving.

On Monday morning, 28 November, when the full team of regular doctors returned from vacation, they insisted we move him right away to a larger, more capable hospital. He was loaded into an ambulance to Fairfax Hospital, and we followed behind in my car. The doctors briefed us that "he might not live through the transfer." While this was shocking, my father had always lived his life on the edge, so we were still under the belief that he would make it.

We pulled into the hospital, and he was taken away through the emergency entrance. Fairfax is a huge hospital; it's the largest hospital in northern Virginia. I said to myself, "This is where he should have been all along." We'd lost four critical days of expert medical support.

Later that day, I called NASA Headquarters and spoke to some of the medical staff there for advice. They didn't have much to offer. The doctors told us that pancreatitis was tough to beat. Our family attitude remained in denial. We'd say, "Look, he has lived his whole life in the one-percentile zone; he will beat it." We simply couldn't get our heads around this force of nature dying.

My mother and sister were "sleeping" in my bedroom. I was on the couch in my media room. My brothers were sleeping on cushions lying on the floor. Jack then flew back to continue with Weapons School. A military family has obligations.

Each day, we would visit the hospital. Around 8 December the doctors made the decision to put him into a coma. We would sit and visit with him, hold his hand, and talk to him.

On the morning of 12 December my mother and sister got up early and were getting ready to head to the hospital. They had just walked out the door when the telephone rang at 6:00 a.m. The voice told me my father had just passed away. I could hear the heart monitor in the background flatlining. I repeated the message to make sure I got it right: "We lost him?" "Yes," they confirmed. I immediately ran to the door and yelled for my mother and sister to come back. By now, my brother Allen had heard me. We all walked out on my balcony. With the sun rising, we cried and hugged. We saw the contrail of an airplane passing overhead. "Daddy is off on the dawn patrol."

My mother then called Jack with the news so that he could make arrangements to fly back to DC. We were just heading out the door to go to the hospital, when the telephone rang. This time, it was the guy who writes obituaries for the *Washington Post*. Clearly, someone on the hospital staff had tipped him off. We all spoke to him, and he wrote his piece.

At the hospital, we gathered his personal effects. I contacted a funeral home near my apartment complex.

I contacted my boss, Dave Clement, the chief of staff of the House Science Committee in the House of Representatives, and requested that Chairman Bob Walker issue a press release. CNN's John Zarrella did a nice piece on my father, as his passing was covered by most of the networks.

We started calling family and friends. For some reason, my mother wanted to speak to Henry Kissinger. Dr. Kissinger had attended my father's launch and was standing next to my father's parents as he walked out of the suit room into the van taking them to the Saturn V. I contacted Alan Shepard's office and relayed the request. Our telephone was ringing off the wall with my friends, my brother's and sister's friends, and my mother's and father's friends, all calling to pass on their condolences. I don't recall how long it took, but at some point, when I answered my telephone, this gravelly voice said, "This is Dr. Kissinger." I replied, "Thank you so much for calling; let me pass you to my mother." My mother was sitting up on the bed. I handed her the phone, walked out, and closed the door.

The superintendent of the cemetery showed us three separate gravesites. In the end, we chose a spot in 7A where many other VIPs are buried. The gravesite is near boxer Joe Louis, actor Lee Marvin, and General Doolittle of Doolittle's Raiders over Tokyo. I picked out a dark marble headstone called Virginia Marble. Since my parents had met in Virginia, I decided it was appropriate. I had an etching done of his Saturn V liftoff and had the stone made large enough that it could also carry an inscription for my mother when she would eventually join my father there.

With the help of our close family friend Colonel Chuck Ikins, we were able to arrange a burial at Arlington National Cemetery for 15 December. That left us only two days to make funeral arrangements. Chuck worked in the Pentagon and had contacts in the Air Force Protocol Office. For a normal military funeral, the official who presents the flag that drapes the casket is equal to or of a higher rank than the person being buried. Thus, my father rated a colonel at a minimum. VIP funerals also take on a larger significance, and members of the honor guard and color guard will sometimes return from leave to be part of it.

At first, when Chuck contacted the Protocol Office about organizing a flyby, he was informed that colonels don't rate a flyby at Arlington Cemetery. That is when Chuck informed the chief protocol officer that this was "no ordinary" colonel; he was the command module pilot of *Apollo 14*. Chuck said, "He looked surprised and said he would start working the issue."

Chuck was amused since the Protocol Office kept calling him for updates on which astronauts were coming. At first, a colonel would present the flag, and then it became a brigadier general. Next, it would be the chief of staff of the U.S. Air Force, the top general. Then it was someone out of the secretary of the air force's office. Finally, it was the deputy secretary of the air force, Rudy DeLeon, who would attend. I am sure each wanted to rub shoulders with the astronauts who would be present.

Most of this time was just a blur. I was emotionally spent and heartbroken that I didn't do more to get my father to a better hospital quicker. I still carry that guilt.

Daddy would tell us, "At my funeral, I want to have the lowest flyby ever. Have them knock the tent down." This was in reference to a funeral where a tent was set up over the gravesite and the missing man formation came so low that the jet wash blew over the tent. The missing man formation is an aerial salute consisting of four airplanes. Above the gravesite, the number-three aircraft pulls hard and zooms vertically skyward, representing the missing pilot on his way to heaven, and the other three aircraft stay in formation, now with a missing aircraft.

My brother Jack, who was an air force fighter pilot, learned which squadron would conduct the missing man formation and spoke to the lead pilot. He transmitted my father's request, and the pilot laughed, "I wish I could, but I can't. We have to be at five hundred feet." He explained that for a flyby at Arlington Cemetery, first they close National Airport from commercial takeoffs, because the formation has to fly directly over the airport. Second, it is the most closely watched radar airspace in the country due to being on the radars of National Airport, Dulles Airport, and Andrews Air Force Base, all of which monitor the skies above the White House, Pentagon, and greater Washington DC. We heard later that as tribute to my father, he led the formation at 499 feet.

We obtained my father's uniform, and on it were his astronaut flight wings. He had previously said that he wanted those to go to Jack. I contacted the NASA Astronaut Office and asked if they could get me a new pair of wings. They said they would and told me there'd be no need to reimburse them. Again, the NASA family supports the smallest of things.

Daddy said there were two things he learned in the U.S. Air Force: how to fly and how to drink beer. In 1981 he was awarded a Coors Beer distributorship for the Mississippi Gulf Coast. At some point, he purchased a Cessna 172. "I am now happy," he'd say, "I am back to flying and drinking beer."

On 14 December we went and conducted a viewing of my father in his casket. He was decked out in his uniform; I placed the new astronaut wings on his chest. I took a Coors Light and placed it next to his right hand. We all told him we loved him.

The one thing that could have stopped the flyby would have been the weather. The day of 15 December was forecast for morning fog and rain all day. That morning, I walked out on the apartment balcony and prayed into the rain, "Daddy, we need you to help out here." We all dressed and stepped into the limo for the short ride to the Old Post Chapel, located at Fort Myer. Fort Myer provides the military support to Arlington Cemetery, everything from the honor guard for the Tomb of the Unknown Soldier to military bands, firing details, marching formations, and the caisson that carries the casket to the church and then to the gravesite.

In the church, we had a section roped off for astronauts, and Alan Shepard delivered the eulogy. Alan Shepard, the first American in space; the commander of my father's mission, *Apollo 14*; and the fifth human being to walk on the moon. The church was packed.

Chuck Ikins leaned over and whispered to his wife, Debbie, "That is something you don't see every day—twenty astronauts in one spot." Senator John Glenn was there, the first American to orbit Earth. There was Neil Armstrong, the first man to walk on the moon. Fellow *14* crewmate Ed Mitchell, the lunar module pilot, was there. On very short notice, the Apollo astronaut corps really turned out. We had at least one representative from each Apollo mission, except for *Apollo 12*, who sent a large funeral wreath with a sash saying it was from "The Crew of *Apollo 12*."

Outside, it was dreary and damp as we exited the limo. Our family walked down and sat on the left-side front bench. All the kids were due to speak before Alan Shepard. I was emotionally drained, so it didn't really hit me as to why we were in the church until I started hearing the noises from outside the church. A booming voice from the parking lot echoed into the church.

"Detail, halt." The sound of boots coming together.

The sound of horse hooves pounded the pavement.

"Present arms." The sound of rifles slapped to attention.

"Order arms." The sound of rifles banged.

I turned and looked at the back of the church. The doors opened as air force pallbearers marched down the aisle with the casket. I had asked my local priest, Father Tuck, to conduct the ceremony. Charlie Duke, my father's good friend and the tenth man to walk on the moon, did the readings.

When my turn came, I walked to the altar, and standing there in my U.S. Marine Corps blues with medals, I said to my father, "Just as you saluted your father at his funeral, I now stand and salute you." I moved my hand slowly up to my cover and then slowly moved it back down to stand at attention. I was emotionally spent.

My brother Jack followed in his U.S. Air Force uniform and read the poem "High Flight," by John Gillespie Magee Jr.

> Oh! I have slipped the surly bonds of Earth
> And danced the skies on laughter-silvered wings;
> Sunward I've climbed, and joined the tumbling mirth
> Of sun-split clouds,—and done a hundred things
> You have not dreamed of—wheeled and soared and swung
> High in the sunlit silence. Hov'ring there,
> I've chased the shouting wind along, and flung

My eager craft through footless halls of air . . .
Up, up the long, delirious burning blue
I've topped the wind-swept heights with easy grace
Where never lark, or ever eagle flew—
And, while with silent, lifting mind I've trod
The high untrespassed sanctity of space,
Put out my hand, and touched the face of God.

Allen and Rosemary told stories of Daddy. Then Alan Shepard spoke of my father and what a good pilot he was. "He used just a thimbleful of fuel to turn the command module around and prepare to dock with the LEM." He measured it off with his thumb against his pinkie finger.

The ceremony ended, and we followed the casket up the aisle and walked outside. The rain was gone, and the sun was shining. The weather was perfect for a flyby.

The air force pallbearers placed the casket on the caisson. The command was given: "Forward march!" The band started to play as the color guard carrying the flag, the firing detail, and a company of U.S. Air Force airman all marched in step as we and the crowd followed. My mother then got into the limousine. She was not able to walk the half mile from the church to the gravesite.

I walked in silence behind the caisson, listening to the band. The weather was cool and crisp. You could hear the hooves of the horses pounding the pavement and pulling the caisson. One of the horses had a saddle, along with a pair of riding boots inserted backward into the stirrups. This is the military's way of symbolizing the missing man or missing rider.

Once at the site, I escorted my mother up to her seat. Neil Armstrong walked up to my mother and said, "Joan, I am not going to stay for the graveside service; I give you my condolences here," and gave my mother a kiss on the cheek. It was as if Neil didn't want to be a distraction to the ceremony. All the other astronauts started to gather around behind the family.

As Chuck and his wife, Debbie, were walking up to the grave site, Chuck whispered marine style (i.e., too loudly) to Debbie, "That is something you don't see every day—the first man on the moon just walking by himself." Neil heard him, looked up, and smiled at Debbie.

Lots more commands:

"Present arms!"

The casket moved into position.

"Order arms!"

At that precise moment, four F-16 jets came streaking past. And the third aircraft pulled up hard into the clear blue sky. As I watched it, I said, "You did it, Daddy. You got us the weather; we got you the flyby."

Then, with the commands of "Ready, Aim, Fire," seven air force members of the firing squad pulled their triggers. Blam! It was heard across the cemetery. Again, "Ready, Aim, Fire." Blam! "Ready, Aim, Fire." Blam! The twenty-one-gun salute was complete.

Next was the playing of "Taps."

The pallbearers folded the flag and gave it to the deputy secretary of the air force. "On behalf of the president of the United States, the United States Air Force, and a grateful nation," he said to my mother, "please accept this flag as a symbol of our appreciation for your loved one's honorable and faithful service." My mother took the flag and placed it in her lap.

Father Tuck then did the readings and concluded the ceremony. I went over and kneeled next to the casket, placing a rose on it. I kissed it and told Daddy I loved him. Each person in my family did something similar.

We had organized a reception at the Fort Myer officers' club. Most of the astronauts came, along with many guests and family members. One thing you learn is that no astronaut fighter pilot is going to admit that someone is a better pilot than he is. That is why he got to be an astronaut, because he believed he was the best. As I made the rounds, shaking their hands and thanking them for coming on short notice, they would say to me, "Just so you know, your father was a damn good pilot." That was the highest compliment they could have given him, and my stick-and-rudder father would have been pleased.

The next day, we loaded up the van and started the lonely drive back home to Gulfport. I was there a few days, cleaning out my father's closet, packing up his personal effects, and getting my mother settled in the house. As a U.S. Marine Corps colonel, it had always struck me how hard it must be each year for families who lose their loved one in active service on a birthday or significant public holiday. Perhaps with this idea in my mind, the following year, my mother started a tradition that she would take the family on a trip shortly before Christmas, just so we were not at home lingering over the memory of his passing away.

Epilogue

I am often asked about my experiences as a child of the Apollo program. On the one hand, we experienced some of the most exhilarating moments in our nation's history and arguably in the history of global exploration; we met world leaders, celebrities, and men and women who pushed the boundary of the possible.

On the other hand, being the child of an astronaut came with expectations and assumptions. Were we gifted? Could we meet or surpass our fathers' legacies? If we did "make it," was it because of our genes or because of some unfair advantage that we had by virtue of being the children of these 1960s heroes? I was a small kid. In fact, when I graduated from high school, I was the second-shortest in my year, and I did not have a large build. It was only after college and joining the U.S. Marine Corps that I grew. My mother assured me that this was something in our genes, but it meant that I was drawn into any number of fistfights at school by some bully who wanted to prove himself against the son of an astronaut.

I suppose the assumption is that I would be gifted with all the same talents as my father and receive advantages because of his status. Certainly, we were able to have some incredible experiences as a result of his work and meet some very talented people, but there were also some expectations that came with the role of being the son of an astronaut that made life difficult for a school-age child.

For one thing, every achievement I had in my early life would be explained away by others by the fact that my father was an astronaut. If I won a ribbon at a track meet, it was not that I had trained hard; it was apparently because my father was an astronaut. If I succeeded in winning a part in the school play, again, apparently, it had nothing to do with my audition and everything to do with being the son of an astronaut. I imagine that this is the same for children of high-achieving athletes or businesspeople.

At some point, as a young man, I decided that I would not talk about my father for this reason. Inevitably, there would be someone who would ask, "There is an Apollo astronaut with your same last name—any relation?" I would say, "Yes, he is my father." Then it would spread like wildfire. After my father passed, it was easier for me to talk about him to others. In part, this was because there was no longer the same expectation that he was assisting me in my every success. But it was also a way to celebrate his life and his remarkable achievements.

Now, as I write this story, it is the occasion of the fiftieth anniversary of my father's mission, *Apollo 14*. I look back at those days of my childhood now and fully appreciate how unique they truly were.

First and foremost, the Apollo astronauts were patriots before that term began to be used as a political label. They had an incredible love of country and honored an intense duty to it and to each other. Many had fought in World War II or Korea or both or could have been sent into combat.

During the Cold War, my father was stationed in Europe, and he was assigned a Polish manufacturing plant as a nuclear target. The plan would have been for him to streak toward the target, pull up, and loft the nuclear bomb into the target before turning away and flying as far away as possible. Everyone involved, from the mission planners to the pilots, knew there was not enough fuel to make it back to home base. He was expected to run out of fuel, try to find an area to bail out, and figure out a way to get back to friendly lines. He knew it was going to be a one-way trip. But if that was what his country needed him to do, he was going to do it.

They had all trained in fighter aircraft, and all wanted to become aces, which requires shooting down five enemy airplanes. The space race was a real race to them. They were determined to beat the Soviets. They were proud to be part of a program that planted the American flag on the moon. My father would often close out his speeches with the following reflection: "Go out and look at the moon tonight. There are six American flags flying on the lunar surface. No other country can make that claim. Be proud of what we as a nation have accomplished."

I grew up amid these incredibly brave men. Almost all of them became great fighter test pilots, men who pushed aircraft to the brink to test the limits of their aircraft. While others had crashed and died, these individuals had survived. Then they became astronauts, destined to fly on rockets built by the lowest bidder.

While the astronauts were full of bravado and humor, they were largely and surprisingly very humble men, from humble beginnings. Most came from small-town USA and just wanted to go fly airplanes. They were born at a unique time in history; they studied and trained hard; and they were the right age to join the military and to be awarded their pilot's wings, distinguish themselves, and apply to the new National Aeronautics and Space Administration. With this confluence of events, they ended up on a rocket going to the moon.

Were it not for the bold challenge issued by a young President Kennedy to put a man on the moon by the end of 1969, despite his tragic assassination, it's not clear the United States would have worked so hard to bring the Apollo program to realization in the time it did.

My father instilled in me this love of country, and my brothers and I have each served as officers in the U.S. military. I went on to become a colonel in the U.S. Marine Corps, because of my pilot-incapable vision; Jack attended the Air Force Academy and flew F-16s, ending as a squadron commander; and Allen went to West Point and became a U.S. Army tank commander.

From an early age, the Apollo astronauts were single-minded about flying. As a child, my father saw two pilots walking the streets of Claremore in their fighter jackets. "At the age of seven, I knew I wanted to be pilot," he would say. This was a common sentiment among the Apollo astronauts. They were the kids building model airplanes in their home and dreaming of flight. I've often found this with other successful people; there's a single-minded pursuit of an interest at a young age, whether it's flying, music, acting, or business. I never had that gift of a singular vision. My interests were always more diffuse and varied, and I wonder if there's a certain kind of success that comes from being solely and passionately focused on a singular area of interest.

The astronauts were also all blessed with extremely good hand-eye coordination. Many were superb hunters and great rifle and shotgun marksmen. Most played golf and were good at it.

Clearly, they were smart. They had to understand engineering, orbital mechanics, computers, geology, aviation, rocket design, and the list goes on and on. Daddy would tell a story about being on a foreign trip with several astronauts when a brassy, loud individual came up to astronaut Bill Pogue and said, "So, Bill, what do you know that I don't know?" His reply was, "A lot." The man never understood the insult. Daddy would always laugh telling the story.

They were all less than six feet tall. Any taller and you couldn't fit in the early capsules. So while they were not physically intimidating by height, they all carried this cocksure aura of self-confidence and talent.

My father was thirty-seven when he went to the moon, and many of the Apollo astronauts were around the same age. As I passed through the age of thirty-seven, I knew that nothing I accomplished could match what my father did at that age by going to the moon. In retrospect, how could I? This was a unique and glorious moment in our nation's history, a defining time for humanity as a whole and a testament to science and the human spirit of curiosity and exploration.

Imagine you are one of the Apollo astronauts. In their midthirties, they undertook the unique work that defined the remainder of their rich lives. From then on, for the rest of their lives, they were the coolest guys in the room, except when Neil Armstrong was around. Who is going to top that? I watched presidents humbling themselves to meet Neil Armstrong and the other astronauts at White House functions. I watched famous musicians, actors, and celebrities stand in awe of them. It seemed the more successful the person, the more in admiration they were of the accomplishments of the astronauts.

My father used to have a pile of fan mail on his office floor. Every so often, he would go through it. His first rule was no self-addressed envelope, no response. As he would say, "Why should I spend my own money to mail my autograph to someone?" Later, when computers became more prevalent, he got tired of seeing something he had just signed turn up for sale on the internet.

In his mind, the men of the astronaut corps were different to other celebrities. Unlike politicians, musicians, and actors, none of the astronauts wanted to become famous. They just wanted to fly airplanes as fast as they could and serve their country; they never sought the limelight. History just put them there.

Before Neil Armstrong died in 2012, I would make the argument that the most famous person in the history of our planet was still walking on it. Some people would say it's all relative, that it depends on which culture you're talking about. So for the Chinese, the Tang dynasty was the golden age of imperial China, and a billion or more Chinese may well nominate an emperor or artist from that time as the most famous person. The same goes for Western cultures, whether it is Alexander the Great, Julius Caesar, Beethoven, William Shakespeare, or George Washington. The list goes on and on. But to my mind, Neil Armstrong was the first person in the history of mankind to walk on the moon, and that is an achievement that transcends any nation or culture. It is an achievement in every history book across the world and will be forever.

The older I get, the more I realize what a special time we lived in. We had a front row seat to the greatest exploration the world had ever seen. The famous people we got to meet as kids. The VIP travel we had on private jets and travel to exotic locations. Hunting trips with men who had walked on the moon. Watching these men lift off to the moon and seeing on TV the nation being torn apart by Vietnam at the same time—go figure.

People would ask my father, "Why are we going into space when we have poor people here on Earth?" He was always perplexed by the question, and I know to him it was a stupid question. He would say, "What do you think happened to the money? Did we put it on a rocket and send it into the sun to burn up? No, it was all spent here on Earth, employing scientists, engineers, technicians, welders and providing construction jobs. It was spent developing technologies we now freely use every day to make our lives better."

In Claremore, Oklahoma, the small town my father grew up in, there is statue to Will Rogers. He was born near Claremore. Rogers was a famous actor, vaudeville performer, cowboy, humorist, and newspaper columnist. There is a large statue of him dressed as a cowboy, with the inscription "I never met a man I didn't like." We would laugh that they should put a statue of Daddy next to it, with him in his space suit and the inscription "Most people are stupid."

My father taught us some valuable life lessons. One of my father's favorite expressions, for example, was, "The harder you work, the luckier you get." My father was about hard work. Another expression of his was, "If you really work eight hours in a day, not spend it joking around the watercooler, wandering the halls, and taking long breaks, you would be surprised at how much you can accomplish in eight hours."

Another of his expressions was, "Have at it." He never told us what to do; but if we did choose to do something, we were expected to work hard at it. My mother would be the one to say, "Stu, that sounds pretty risky. Why don't you take a look at it?" As a young kid, I remember one time a neighboring kid and I built a rinky-dink little raft made out of two-by-four studs and some plywood. Daddy suggested we put two tire inner tubes inside of it to help it float. We took it to Taylor Lake and launched it. It didn't work that well. We then came up with the idea of building a submarine. Daddy was all gung ho, but I think my mother talked us out of the idea. Common sense prevailed.

He was a man of action. One simple memory here stands out. Around our house in El Lago, Texas, there were many fields and woods. When we were teenagers, we had dirt bikes, and we would all ride the trails. There was a popular spot that had a big open ditch where you could ride and fly out of the ditch and get some airtime. As we got better and flew farther each time, a twenty-foot pine tree started to become a hazard. One night, we were complaining about the tree. Without hesitation, Daddy drove my brothers and I in our Chevy Blazer to the ditch. Once there, he pulled out a chain saw and, in about two minutes, took down that tree. The next day, we returned to ride and saw all the other dirt bike riders excited by the fact the tree was gone. We admitted to nothing, but I took pride knowing my father had made that possible for all of us, because my father took action.

Most of all, my father inspired me. He would tell us the story of his youth and his journey to NASA: "If a poor kid can go to the moon, you can do anything. Just set your mind to it and work hard at it."

On life, he explained to us, "No matter how difficult an issue you are facing, a year from now, you will wonder why you were so worried about it. I can guarantee you this: twenty years later you will look back and laugh that you got so worked up about it."

To be fair to my mother, my parents were a great team. My mother played the game with panache and joy and helped his career. My mother loved life. She taught us to dance, to cook, to travel, to study history, and to have proper etiquette for any situation in life. My mother passed away in 2007.

Now, as the father of five young children, I wonder what legacy I will leave for my kids. Have I made a mark on history? Certainly not as big as my father. What am I giving them a front row view to? Am I giving them the life skills they need to face the challenges of the future with confidence, intelligence, and curiosity? While I can't predict the future, I will always teach them the lessons of their grandfather and grandmother.

I will forever be a son of Apollo.

In the Outward Odyssey: A People's History of Spaceflight series

Into That Silent Sea: Trailblazers of the Space Era, 1961–1965
Francis French and Colin Burgess
Foreword by Paul Haney

In the Shadow of the Moon: A Challenging Journey to Tranquility, 1965–1969
Francis French and Colin Burgess
Foreword by Walter Cunningham

To a Distant Day: The Rocket Pioneers
Chris Gainor
Foreword by Alfred Worden

Homesteading Space: The Skylab Story
David Hitt, Owen Garriott, and Joe Kerwin
Foreword by Homer Hickam

Ambassadors from Earth: Pioneering Explorations with Unmanned Spacecraft
Jay Gallentine

Footprints in the Dust: The Epic Voyages of Apollo, 1969–1975
Edited by Colin Burgess
Foreword by Richard F. Gordon

Realizing Tomorrow: The Path to Private Spaceflight
Chris Dubbs and Emeline Paat-Dahlstrom
Foreword by Charles D. Walker

The X-15 Rocket Plane: Flying the First Wings into Space
Michelle Evans
Foreword by Joe H. Engle

Wheels Stop: The Tragedies and Triumphs of the Space Shuttle Program, 1986–2011
Rick Houston
Foreword by Jerry Ross

Bold They Rise: The Space Shuttle Early Years, 1972–1986
David Hitt and Heather R. Smith
Foreword by Bob Crippen

Go, Flight! The Unsung Heroes of Mission Control, 1965–1992
Rick Houston and Milt Heflin
Foreword by John Aaron

Infinity Beckoned: Adventuring Through the Inner Solar System, 1969–1989
Jay Gallentine
Foreword by Bobak Ferdowsi

Fallen Astronauts: Heroes Who Died Reaching for the Moon, Revised Edition
Colin Burgess and Kate Doolan with Bert Vis
Foreword by Eugene A. Cernan

Apollo Pilot: The Memoir of Astronaut Donn Eisele
Donn Eisele
Edited and with a foreword by Francis French
Afterword by Susie Eisele Black

Outposts on the Frontier: A Fifty-Year History of Space Stations
Jay Chladek
Foreword by Clayton C. Anderson

Come Fly with Us: NASA's Payload Specialist Program
Melvin Croft and John Youskauskas
Foreword by Don Thomas

Shattered Dreams: The Lost and Canceled Space Missions
Colin Burgess
Foreword by Don Thomas

The Ultimate Engineer: The Remarkable Life of NASA's Visionary Leader George M. Low
Richard Jurek
Foreword by Gerald D. Griffin

Beyond Blue Skies: The Rocket Plane Programs That Led to the Space Age
Chris Petty
Foreword by Dennis R. Jenkins

A Long Voyage to the Moon: The Life of Naval Aviator and Apollo 17
Astronaut Ron Evans
Geoffrey Bowman
Foreword by Jack Lousma

The Light of Earth: Reflections on a Life in Space
Al Worden with Francis French
Foreword by Dee O'Hara

Son of Apollo: The Adventures of a Boy Whose Father Went to the Moon
Christopher A. Roosa
Foreword by Jim Lovell

To order or obtain more information on these or other
University of Nebraska Press titles, visit nebraskapress.unl.edu.